FUNDAMENTALS OF
FLUID MECHANICS

FUNDAMENTALS OF FLUID MECHANICS

James A. Sullivan
Southern Illinois University
Carbondale, Illinois

RESTON PUBLISHING COMPANY, INC.
Reston, Virginia
A Prentice-Hall Company

Library of Congress Cataloging in Publication Data

Sullivan, James A
    Fundamentals of fluid mechanics.

    Includes index.
    1. Fluid mechanics. I. Title.
TA357.S86   620.1'06   78-696
ISBN 0-8359-2999-X

10 9 8 7 6 5 4 3 2 1

Printed in the United States of America.

To Sylvia

CONTENTS

PREFACE

Fluid mechanics[1] is an applied study with much of its basis in experiment. It is concerned with the motion of matter that flows or tends to flow. It combines applied mathematics and physics that describe particle motions or tendencies to motion, with those special aspects of fluids that permit them to flow or govern their motion. Fluids may be liquids or gases. That branch of fluid mechanics which is concerned with the liquids water and oil is called *hydromechanics*. Hydromechanics, in turn, is divided into hydrostatics (liquids at rest), hydrodynamics (liquids in motion), and hydraulics (compressed liquids in pipes). That branch of fluid mechanics which is concerned with air and gases is called *aeromechanics*, which can be subdivided into aerostatics (air and gases at rest), aerodynamics (air and gases in motion), and aeromatics (compressed air and gases in pipes), which is commonly called pneumatics.

Fundamentals of Fluid Mechanics seeks to correlate the theoretical framework and concepts that describe the behavior of fluids with practical applications and work in laboratories, testing facilities, and shops. Suggested learning and performance are intended to prepare persons to assume technical positions that assist engineering and maintenance projects, or to work with equipment that processes fluids in the generation and use of power. Many of the suggested activities have been derived from laboratory exercises and tasks commonly assigned to technicians by industry. Others are taken from ASTM standardized procedures.

The International System of Units (SI), which are absolute, and English units, which are gravitational, are used throughout this text.

---

[1]G. A. Tokaty, *A History and Philosophy of Fluidmechanics.* Haney-on-Thames, Oxfordshire, England: G. T. Foulis and Co. Ltd., 1973.

Although SI units are given preference, except in the case of reporting derivations with origins in the English system, it is also recognized that English units will be used for some time during and even after the transition period when SI is being adopted. Quantities and units in the English system are inserted in brackets immediately following the SI designated quantities in the examples. Equivalent answers to the example problems are also bracketed in English units following the SI answers. The student with a preference for English units may wish to confirm these by working through the problems with the English units given. The review problems that follow each chapter are stated both in SI and English units, as are the answers in the appendix.

Although it is common practice to introduce technicians to calculus during the course of their preparation, it is not seen as necessary to make use of this method in the explanation and solution of problems in a first course in fluid mechanics. It is necessary, however, to be familiar with the rudiments of algebra, as these are used extensively to derive concepts that describe the behavior of fluids and to solve related problems.

Many persons and companies have assisted the author in the preparation of this work. Special thanks is extended to several reviewers for their advice, including Tobi Goldoftas and Chris R. Treleaven. K. Gita Balagopalon and S. B. Surish are acknowledged for proofing and working sample problems. Finally, Sedat Sami is acknowledged for his counsel in the subject.

James A. Sullivan

1

DIMENSIONS
OF FLUID MECHANICS

1-1 INTRODUCTION

Fluid mechanics considers the behavior of fluids at rest and in motion, in open and closed systems. Fluids at rest are treated in the study of hydrostatics and aerostatics, whereas fluids in motion are treated in the study of hydrodynamics, hydraulics, and aeronautics. Liquids and gases are both considered to be fluids. Open systems contain fluids in conveyances open to the atmosphere, whereas closed systems confine liquids and gases in pipes, typically under pressure. Hydraulics, which considers the flow and control of liquids in pipes, is an example of a closed system.

When fluids are in equilibrium—that is, in a balanced condition—they cannot sustain a shear stress. This is the case when a fluid is not flowing. Shear stresses are those forces acting in opposite directions at a tangent to adjacent surfaces. A bolt such as that in Fig. 1-1, for example, used to hold two plates together, is withstanding a shear stress when the forces are acting in opposite directions at a tangent. If the outer surface of the bolt were imaginary, and water were substituted for the material within the bolt, the plates would move, indicating that the water could not sustain the stress required to keep the plates from slipping.

Fluids in equilibrium cannot withstand shear stresses because of the loose arrangement of the molecules within the substance. The spaces between the molecules are large and other cohesive forces between the molecules are small, permitting considerable freedom of movement between the molecules. The difference between liquids and gases can be defined in part from the relative spacing of the molecules. This is much larger in the case of gases, contributing to the reason that they flow more readily than liquids. The larger spacing in gases also accounts for the

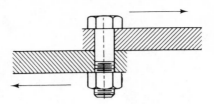

Fig. 1-1 Shear stress in bold fastener

greater compressibility of gases as compared with liquids.

Solutions to problems in applied fluid mechanics are derived from data secured from instruments that monitor system performance, as well as from a knowledge of certain properties of the fluids themselves. Pressure and flow rate, for example, are monitored with instruments attached directly to operational systems, whereas fluid properties such as density and viscosity are determined beforehand in the laboratory under controlled conditions. In both cases, however, units for the quantitative measurement must be established which make the available data intelligible. That is, a uniform system of units must be adopted so that both the data used in computing solutions to fluid mechanics problems and the solutions themselves will be understood and useful to people who are concerned. Until recently, both English and metric units have been used in computing the solutions to fluid mechanics problems, causing the student considerable difficulty in making the transition between the two systems and creating confusion when the same units have two apparent meanings. Use of the kilogram to measure both force and mass is an excellent example of a conflicting definition that causes confusion.

1-2 HISTORICAL PERSPECTIVE

Metric measurement has received widespread use for several hundred years in the sciences in America and abroad. Recent emphasis by international bodies to standardize metrication for all countries has resulted in agreement on a practical system of units of measurement termed in French the Système International d'Unités (International System of Units), with SI being the official abbreviation.

There are three classes of SI units: base units, derived units, and supplementary units.[1] Seven base units have been designated for length, mass, time, electric current, thermodynamic temperature, amount of substance, and luminous intensity (Table 1-1). Derived units are formed by

---

[1]*The International System of Units (SI)*, National Bureau of Standards, U.S. Department of Commerce, NBS Special Publication 330 (Washington, D.C.: U.S. Government Printing Office, 1971), pp. 1–17.

combining base units according to the algebraic relations linking the corresponding quantities (Table 1-2). Several of these algebraic expressions in terms of base units can be replaced by special names and symbols which can themselves be used to form other derived units. Table 1-3 lists several examples of SI derived units expressed by means of special names. Supplementary units are those that the General Conference of Weights and

TABLE 1-1   SI base units

Quantity	Name	SI Symbol
length	meter	m
mass	kilogram	kg
time	second	s
electric current	ampere	A
thermodynamic temperature	kelvin	K
luminous intensity	candela	cd
amount of substance	mole	mol

TABLE 1-2   SI derived units expressed in terms of base units

Quantity	Name	SI Symbol
area	square meter	m^2
volume	cubic meter	m^3
speed, velocity	meter per second	m/s
acceleration	meter per second squared	m/s^2
density, mass density	kilogram per cubic meter	kg/m^3
concentration (of amount of substance)	mole per cubic meter	mol/m^3
specific volume	cubic meter per kilogram	m^3/kg
luminance	candela per square meter	cd/m^2

TABLE 1-3   Examples of SI derived units expressed by means of special names

Quantity	SI Units		
	Name	Symbol	Expression in terms of SI base units
force	newton	N	$m \cdot kg \cdot s^{-2}$
pressure	pascal[7]	Pa	$m^{-1} \cdot kg \cdot s^{-2}$
energy, work, quantity of heat	joule	J	$m^2 \cdot kg \cdot s^{-2}$
power, radiant flux	watt	W	$m^2 \cdot kg \cdot s^{-3}$
quantity of electricity, electrical charge	coulomb	C	$s \cdot A$
dynamic viscosity	pascal second	Pa·s	$m^{-1} \cdot kg \cdot s^{-1}$
moment of force	newton meter	N·m	$m^2 \cdot kg \cdot s^{-2}$
surface tension	newton per meter	N/m	$kg \cdot s^{-2}$
specific heat capacity, specific entropy	joule per kilogram kelvin	J/(kg·K)	$m^2 \cdot s^{-2} \cdot K^{-1}$
specific energy	joule per kilogram	J/kg	$m^2 \cdot s^{-2}$
energy density	joule per cubic meter	J/m^3	$m^{-1} \cdot kg \cdot s^{-2}$

Measures (CGPM) has not yet classified as base or derived units of the International System. Two supplementary units, the plane angle, or radian, the solid angle, or steradian, have been identified.

1-3   THE INTERNATIONAL SYSTEM OF UNITS

The unit of length illustrated in Fig. 1-2 is the meter. It is defined as being equal to 1 650 763.73 wavelengths in vacuum of the radiation corresponding to the transition between the levels $2p_{10}$ and $5d_5$ of the krypton-86 atom.[2] The previous international prototype of platinum-iridium, which was legalized by the first CGPM is kept at the International Bureau of Weights and Measures (BIPM) in Paris, France under the conditions specified in 1889.

The unit of mass shown in Fig. 1-3 is the kilogram; it conforms to the prototype made of platinum-iridium kept at the BIPM under the conditions specified by the first CGPM.[3] A duplicate in the National Bureau of Standards, U.S. prototype kilogram 20, serves as the mass standard for the United States. This is the only base unit still defined by an artifact. Mass is not weight or force; rather, it is the molecular amount of a substance.

The unit of time is the second. Originally, the second was defined as a fraction of the solar day, but the definition was changed because measurements have shown that because of irregularities in the rotation of the earth the mean solar day does not guarantee the desired accuracy. The thirteenth CGPM (1967) passed a resolution that defined the second as the duration of 9 192 631 770 periods of the radiation corresponding to the transition between the two hyperfine levels of the ground state of the cesium-133 atom (Fig. 1-4).

The unit of electrical current illustrated in Fig. 1-5 is the ampere; it is defined as that constant current which, if maintained in two straight parallel conductors of infinite length, of negligible circular cross section, and placed one meter apart in vacuum, would produce between these conductors a force equal to 2×10^{-7} newton per meter of length.[4]

The unit of thermodynamic temperature is the kelvin and is defined as the fraction $1/273.16$ of the thermodynamic temperature of the triple point of water.[5] In addition to the thermodynamic temperature expressed in kelvins (symbol T), use is also made of Celsius temperature (symbol t), defined by the equation

$$t = T - T_0$$

[2]11th CGPM (1960), Resolution 6.
[3]Third CGPM (1901).
[4]Ninth CGPM (1948).
[5]Thirteenth CGPM (1967), Resolution 3 and Resolution 4.

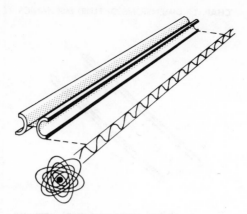

Fig. 1-2 Unit of length—meter

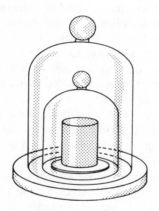

Fig. 1-3 Unit of mass—kilogram

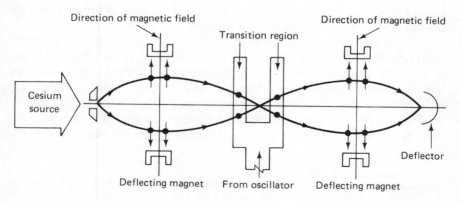

Fig. 1-4 Unit of time—second

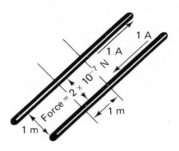

Fig. 1-5   Unit of electrical current—ampere

where $T_0 = 273.15°K$ by definition. The unit "degree Celsius" is thus equal to the unit "kelvin" and an interval of difference of Celsius temperature may also be expressed in degrees Celsius.

The unit of luminous intensity, the candela, is measured in the perpendicular direction on the surface of $1/600,000$ square meter of a blackbody at the temperature of freezing platinum under a pressure of 101,325 newtons per square meter.[6]

The unit of a substance is the mole, defined as the amount of a substance of a system which contains as many elementary entities as there are atoms in 0.012 kilogram of carbon 12. When the mole is used, the elementary entities must be specified and may be atoms, molecules, ions, electrons, other particles, or specified groups of such particles.

The supplementary unit, the radian, is the plane angle between two radii of a circle which cut off on the circumference an arc equal in length to the radius. The supplementary unit, the steradian, is the solid angle which, having its vertex in the center of a sphere, cuts off an area of the surface of the sphere equal to that of a square with sides of length equal to the radius of the sphere. These are illustrated in Fig. 1-6.

The International Organization for Standardization (ISO) has issued additional recommendations with the aim of securing uniformity in the use of units, in particular those of the International System.[7] According to these recommendations, the product of two or more units is preferably indicated by a dot. The dot may be dispensed with when there is no risk of confusion with another unit symbol, for example, Nm or N·m, but not mN. A solidus (oblique stroke,/), a horizontal line, or negative powers

---

[6]Thirteenth CGPM (1967), Resolution 5.

[7]International Organization for Standardization in publication ISO 2944-1974 (E) defines the unit of pressure as the bar, where 1 bar = 100 kPa $\approx$ 14.5 lbf/in². However, the official designation for pressure in SI units is the Pascal (Pa). Because the Pascal is a small unit, pressure gauges are commonly constructed with dials graduated in units of one thousand Pascals (kPa).

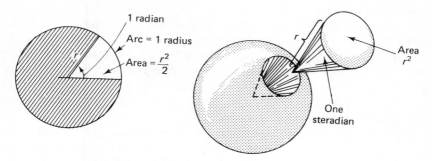

Fig. 1-6 Supplementary units—radian and sterandian

TABLE 1-4 Multiple and submultiple units of the SI System

Prefix	SI Symbol	Multiplication factor
tera	T	$1\ 000\ 000\ 000\ 000 = 10^{12}$
giga	G	$1\ 000\ 000\ 000 = 10^{9}$
mega	M	$1\ 000\ 000 = 10^{6}$
kilo	k	$1\ 000 = 10^{3}$
hecto	h	$100 = 10^{2}$
deka	da	$10 = 10^{1}$
deci	d	$0.1 = 10^{-1}$
centi	c	$0.01 = 10^{-2}$
milli	m	$0.001 = 10^{-3}$
micro	μ	$0.000\ 001 = 10^{-6}$
nano	n	$0.000\ 000\ 001 = 10^{-9}$
pico	p	$0.000\ 000\ 000\ 001 = 10^{-12}$
femto	f	$0.000\ 000\ 000\ 000\ 001 = 10^{-15}$
atto	a	$0.000\ 000\ 000\ 000\ 000\ 001 = 10^{-18}$

may be used to express a derived unit formed from two others by division, for example, m/s, $\frac{m}{s}$, or $m \cdot s^{-1}$. The solidus must not be repeated on the same line unless ambiguity is avoided by parentheses. In complicated cases negative powers or parentheses should be used; for example, m/s^2 or $m \cdot s^{-2}$ is appropriate, as in $mkg/(s^3A)$ or $m \cdot kg \cdot s^{-3}A^{-1}$, but not $m/s/s$ or $m \cdot kg/s^3/A$.

The SI system of units is based on decimal arithmetic, as is the traditional metric system. For each quantity defined, multiplication and division are performed using powers of ten by moving the decimal point to the right or left and adding zeros. For example, the meter equals 1000 millimeters (mm × 1000 = m), and the kilometer equals 1000 meters (km × 0.001 = m).

The abbreviation system for multiples and submultiples of the eight base units[8] in the SI system is shown in Table 1-4. Notice that there are six prefixes for multiples and that mega-, giga-, and tera- use uppercase letters,

---

[8]Eleventh CGPM (1960), Resolution 12.

whereas deca-, hecto-, and kilo- use lowercase letters. The number of zeros past the whole number equals the value of the exponent, for example, $1000 = 10^3$. There are eight prefixes for the submultiples and while none use uppercase letters, notice that the Greek letter mu (μ) is used to denote micro. The number of places past the decimal is indicated by the value of the exponent. The negative exponent indicates that the zeros are placed to the right of the decimal, with the value of the exponent being equal to the number of places past the decimal point.

Multiplication using the SI system is accomplished by multiplying the whole or decimal numbers and adding or subtracting the exponents, in accordance with the basic laws of algebra. For example, for positive exponents

$$(4.6 \times 10^4) \times (3.2 \times 10^7) = 14.72 \times 10^{11} = 15 \times 10^{11}$$

Because both exponents are positive, they are added and the sign remains positive. As another example,

$$(5.9 \times 10^6) \times (2.1 \times 10^{-9}) = 12.39 \times 10^{-3} = 12 \times 10^{-3}$$

Since one exponent is positive and the other is negative, the smaller is subtracted from the larger, and the sign of the larger is retained, which follows the basic law of signs in algebra.

Division using the SI system is accomplished by dividing the whole or decimal numbers, moving the exponent in the denominator into the numerator, changing its sign, and then following the basic law of signs used in multiplication. One example,

$$\frac{4.6 \times 10^4}{3.2 \times 10^7} = 1.4375 \times 10^4 \times 10^{-7} = 1.4 \times 10^{-3}$$

Another example,

$$\frac{5.9 \times 10^6}{2.1 \times 10^{-9}} = 2.8095 \times 10^6 \times 10^9 = 2.8 \times 10^{15}$$

The seven SI base units form a coherent system. A system of units is coherent if, when two unit quantities are multiplied or divided by each other, the resultant quantity is also a unit quantity. For example, where the meter is the unit of length, the square meter is the unit of area. By contrast, were the unit of length to be the foot, the unit of area would be the square foot, not the acre.

Several basic rules have been established to maintain the coherent system. Only SI units and their multiples or submultiples should be used.

These include the base units, supplementary units, derived units, and their combinations. Only the metric units themselves should be used in combination, not their submultiples, to form derived units. This means, for example, that pressure derived as newtons per square meter (N/m^2) is correct, but newtons per square centimeter (N/cm^2) or newtons per square millimeter (N/mm^2) would not be. Prefixes may be used in the numerator of a resulting combination for purposes of clarity, for example, (MN/m^2), but should not appear in the denominator. The units themselves without prefixes should be used when calculations are made. Insignificant digits and decimals are eliminated by using approved prefixes to indicate the order of magnitude of the number, for example, 150000 cm should be written 1500 m or 1.5 km. Calculations using the base unit are usually made by substituting appropriate powers of ten, for example, 1.5×10^3 m in the previous example. Prefixes indicating intervals of 1000 are preferred, i.e., mm, m, km, and Mm. To facilitate the reading of numbers with four or more digits, they should be set off in groupings of three, for example, 1000000 should be written 1 000 000. Commas, which are commonly used as decimals in Europe and elsewhere to set off groups of three digits, should be avoided.

Conversion of units and rounding should be handled carefully. When conversions are made, the complete equivalents should be used in the calculations before rounding so that accuracy is not distorted or sacrificed. Equivalent numbers can be rounded in accordance with Table 1-5. If a number is to be rounded off to fewer digits than the total number available, the following rules should be used.[9] If the first digit discarded is less than 5, the last digit retained should not be changed. For example, 9.012343 rounded to six digits would be 9.01234, to five digits 9.01234, and to four digits 9.012. When the first digit discarded is greater than 5, the first digit retained should be increased by one unit; for example, 8.012678 rounded to six digits would be 8.01268, to five digits 8.0127, and to four digits 8.013. When the first digit discarded is exactly 5 followed only by zeros, the last digit retained should be increased by one unit if it is an odd number, but not adjusted if it is an even number; for example, 5.312545 rounded to six places would be 5.31254.

When digits are rounded off, consideration must be given to the intended accuracy and the significance of the number.[10] A digit is said to

[9]*ASTM Metric Practice Guide*, National Bureau of Standards, U.S. Department of Commerce, Handbook 102 (Washington, D.C.: U.S. Government Printing Office, 1967), pp. 11–13.

[10]See also ASTM Recommended Practices E 29, for Designating Significant Places in Specified Limiting Values, *1966 Book of ASTM Standards*, Part 30.

TABLE 1-5   Rounding of minimum and maximum limits (*Reprinted from* ASTM *Metric Practice Guide,* National Bureau of Standards)

Numerical range		Round to nearest
from	*but less than*	
0.000	0.025	0.0001
0.025	0.05	0.0005
0.05	0.25	0.001
0.25	0.5	0.005
0.5	2.5	0.01
2.5	5	0.05
5	25	0.1
25	50	0.5
50	250	1
250	500	5
500	2 500	10
2 500	5 000	50
5 000	25 000	100
25 000	50 000	500
50 000	250 000	1 000

be significant if it is necessary to define a specific value or quantity. Zeros may indicate either a specific value or the magnitude of the number.

Each of the following list of numbers has a different magnitude, but contains only one significant digit:

$$1000$$
$$100$$
$$10$$
$$1$$
$$0.01$$
$$0.001$$
$$0.0001$$

Numbers drawn from different information sources are frequently added, subtracted, multiplied, and divided, and specific rules must be followed. In addition and subtraction, the answer cannot contain more significant digits to the right than are contained in the least accurate figure. This requires that the numbers that are added or subtracted should first be rounded to one more right-hand significant digit than is contained in the least accurate number. Then the final answer is rounded off. For example,

21 367 268	first is rounded to	21 000 000
38 000 000		38 000 000
378 348		400 000
		59 000 000

and the final answer is rounded off to 59 000 000. In multiplication and division, the product or quotient cannot contain more significant digits than are contained in the number with the fewest significant digits used in the multiplication or division. This means that, if one of the numbers contains three significant digits, the answer, rounded off, is limited to three significant digits, regardless of the placement of the decimal point.

1-4   MASS, WEIGHT, AND ACCELERATION

Mass is defined by the international standard kilogram, which contains molecules of a specific type occupying the available space that constitutes the confines of that prototype. It is not a force or weight. During all ordinary circumstances, the mass of a body remains constant and never changes. Thus, the mass of a body is given and defined by its composition and size.

Weight, on the other hand, results from the gravitational pull of the earth acting on a given mass. Weight is derived in force units, newtons. Whereas force may be in any direction, weight is considered to be in the vertical plane. The weight of a body is affected by changes in gravity. As gravity increases, the weight of the body increases, and, conversely, as gravity decreases, the weight of the body decreases. By definition, force in newtons is that force which, when applied to a body having a mass of one kilogram, gives it an acceleration of one meter per second squared (m/s^2). It is independent of the force of gravity acting on a body (weight), even though both are measured in force units.

The mass and weight of a body are related by Newton's second law of motion. Simply stated, the law says that if a body is acted upon by a force F, the acceleration a is equal to the value of the force divided by the mass M of the body. In notation,

$$a = \frac{F}{M} \tag{1.1}$$

If weight is measured in newtons, and the mass is measured in kilograms, acceleration will be computed in meters per second squared.

The value of acceleration due to gravity is given the symbol g and is taken as 9.80665 m/s^2. Substituting in the general equation $a = F/M$, weight w for force, and the gravitational constant g for the acceleration due to gravity, we obtain

$$\text{Gravitational constant } (g) = \frac{\text{Weight}(w)}{\text{Mass}(M)}$$

and

$$g = \frac{w}{M} \tag{1.2}$$

Solving for w, we obtain

$$w = Mg \tag{1.3}$$

or

newtons (w) = kilograms $(M) \times 9.80665$ meters per second squared (g)

Adoption of the newton as the unit of force is receiving widespread attention from engineers and technicians. English and traditional metric systems use weight units to describe force units such that when applied, they give the unit of mass an acceleration equal to g, given a standard or constant value of 32.174 ft/sec² or 9.80665 m/s² (commonly rounded off to 32.2 ft/sec² or 9.8 m/s²). The newton, on the other hand, is defined as the force that generates a unit acceleration of one meter per second squared to a mass of one kilogram, independent of the gravitational constant g. Consequently, in the SI system, g will appear in many formulas in fluid statics, but disappear from others used in fluid dynamics where it was formerly present. For example, the weight of a mass of M kilograms is equal to a force of Mg newtons, where g is given as 9.806650 m/s².

1-5 INTRODUCTION TO DIMENSIONAL ANALYSIS

Dimensional analysis has application to both theoretical and applied solutions to fluid mechanics problems. It is the mathematics of dimensions. Specifically, it deals with establishing exact relationships between measurement systems, for example, between SI and other systems, as well as assuring the internal consistency within a single numbering system, that is, compatible units.

Measurement systems and equations are homogeneous, that is, internally consistent, only if they are the same physically and are measured in the same units. This is to say that in an equation that expresses a physical relationship, the two sides of the equation not only must be physically equal, but dimensionally equal as well. Equating area to volume, for example, or mixing English and metric units would upset the relationship dimensionally.

In application, dimensional analysis techniques can be used to convert one measurement system to another, develop equations that establish relationships between quantities, reduce the number of variables in relationships to fundamental quantities, and predict full-size prototype behavior from model similitude, the study of scale prototype models in hydraulic research.

So long as different measurement systems are in use, considerable effort will be expended to ensure that, even when equations do express the

TABLE 1-6 Comparison of SI metric and traditional English systems

System	Weight	Mass	Acceleration
SI metric	newton	kilogram	meter per second squared
English	pound	slug	feet per second squared

same physical quantity, consistent units are used throughout. Two common systems and their base units for weight, mass and acceleration are shown in Table 1-6. When information about the physical characteristics of a substance or system is given in units other than SI, conversions must be made to SI units before equations can be made dimensionally homogeneous and solved correctly. A number of sources are available to convert other systems to SI units.[11] Use of Eq. (1.3) and the approximate relationships $N = lb \times 4.448$, $kg = slugs \times 1.459 \times 10^1$, and $meters = ft \times 3.048 \times 10^{-1}$ are used here to illustrate a typical example.[12]

EXAMPLE 1-1

Determine the weight of a substance that has a mass of 5 slugs.

SOLUTION
First convert slugs to kilograms.

$$Kg = 5 \times 14.59 = 72.95$$

Then, using Eq. (1.3),

$$w = Mg$$

and

$$w = (72.95 Kg)(9.8 \ m/s^2) = 714.9 \ N$$

In engineering, dimensional analysis is commonly used to establish relationships between measurement systems and ensure internal consistency by converting all dimensions to the fundamental units of mass, length, and time (MLT); or force, length, and time (FLT). This can be illustrated by using Newton's second law of motion expressed in Eq. (1.1). If mass is the fundamental unit, force can be written dimensionally as

$$Force = mass \times \frac{distance}{time^2}$$

[11]*ASTM Metric Practice Guide, op. cit.*
[12]It should also be remembered that in the traditional metric system, the kilogram force (kgf) receives widespread use.

or symbolically as MLT^{-2}. If force is the fundamental unit, mass can be written dimensionally as

$$\text{Mass} = \text{force} \times \frac{\text{time}^2}{\text{distance}}$$

or symbolically as $FL^{-1}T^2$. In some systems, temperature θ (Greek symbol theta) is added as a fourth dimension.

Work, for example, is measured as distance times force. Kinetic energy, on the other hand, is computed as one-half the product of mass times the speed squared. In MLT (preferred in SI) and FLT units, however, these quantities are mechanically and dimensionally equivalent. That is,

$$\text{Work} = \text{distance} \times \text{force} = FL = (MLT^{-2})(L) = ML^2T^{-2}$$

and

$$\text{Kinetic energy} = \tfrac{1}{2}Mv^2 = (FL^{-1}T^2)(LT^{-1})^2 = FL = ML^2T^{-2}$$

Table 1-7 lists several familiar quantities in MLT and FLT units that are used in solving fluid mechanics problems.

Answers to problems in fluid mechanics are typically computed in dimensional units such as area, pressure, power, and volume flow. Equations are constructed so as to position the desired quantity as the dependent variable. For example, in solving for the volume of the box shown in Fig. 1-7 with equal sides of $x, y,$ and z of meter length,

$$V = x^1y^1z^1 = (1^{1+1+1})(m^{1+1+1}) = 1^3m^3$$

or

$$V = xyz = 1\,m^3$$

TABLE 1-7   Quantities in MLT and FLT units

Quantity	Equation Symbol	MLT units	FLT units
mass	M	M	$FL^{-1}T^2$
force	F	MLT^{-2}	F
pressure	Pa	$ML^{-1}T^{-2}$	FL^{-2}
distance	L	L	L
time	t	T	T
area	A	L^2	L^2
volume	V	L^3	L^3
speed, velocity	v	LT^{-1}	LT^{-1}
acceleration	a	LT^{-2}	LT^{-2}
specific weight	γ	$ML^{-2}T^{-2}$	FL^{-3}
density, mass density	ρ	ML^{-3}	$FL^{-4}T^2$
work	G	ML^2T^{-2}	FL
kinetic energy	KE	ML^2T^{-2}	FL
power, radiant flux	W	ML^2T^{-3}	FLT^{-1}

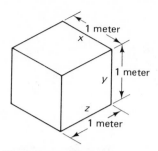

Fig. 1-7 Volume equals 1 m³

where V is the dependent variable, x, y, and z are the independent variables, and m^3 is the dimensional unit with an exponent of 3. The exponent in this case is a dimensionless number and represents the ratio of the volume exponent to the exponent of the average length of the side dimensions $x, y,$ and z. That is,

$$\frac{\text{Volume exponent}}{\text{Average side length exponent}} = C = 3$$

and the constant $C = 3$ in this case, regardless of the size of the cubic container.

Dimensionless numbers have application in solving for the general form of many questions. In the equation for kinetic energy, for example, $KE = \frac{1}{2}mv^2$ can be written dimensionally in MLT units as

$$KE = ML^2T^{-2}$$

and the exponents for M, L, and T are 1, 2, and -2, respectively.

EXAMPLE 1-2

Develop an equation to solve for the distance l traveled by a free-falling body in time t, assuming that distance is a function f of the weight, gravity, and time.

SOLUTION

Since the distance is thought to be a function of three variables,

$$l = f(w, g, t)$$

Both sides of the equation must be equal and homogeneous. Designating the known exponent on the left as 1, and the unknown exponents on the right as a, b, and c, we have

$$l^1 = f(w^a g^b t^c)$$

In *MLT* units

$$M^0L^1T^0 = f(M^aL^aT^{-2a})(L^bT^{-2b})(T^c)$$

Equating exponents on the left side of the equation with those on the right gives

$$0 = a$$
$$1 = a + b$$
$$0 = -2a - 2b + c$$

Solving the first two of these simultaneously, $a = 0$ and $b = 1$. Substituting these values in the third gives $c = 2$, and the derived equation becomes

$$l = f(w^0 g^1 t^2)$$

or

$$l = Cgt^2$$

where $w^0 = 1$ and C is a dimensionless constant that cannot be determined from this dimensional reasoning. It must be determined by experiment.

1-6 SUMMARY

The study of fluid mechanics includes liquids and gases in equilibrium as well as in motion. Open systems are subject to the effects of gravity and atmospheric conditions, whereas closed systems, such as hydraulic systems, which pipe the fluid under high pressure are affected little by these changes. When fluids are in equilibrium, they cannot sustain a shear stress because of their loose molecular arrangement, the magnitude of which also distinguishes liquids from gases.

Worldwide adoption of the International System of Units will reduce difficulty encountered with the several measurement systems currently in use. The seven base units, supplementary units, and derived units form a coherent system. Because the system is based on decimal arithmetic, multiples and submultiples are easily formed by moving the decimal place and using the designated prefix. Conversion from other numbering systems to SI units, rounding, multiplication, and division are performed according to set procedures to maintain the coherence of the system.

Mass, weight, and acceleration are related by using Newton's second law of motion. There is a major difference between the SI force unit and the English force unit. The newton is defined as that force which when applied horizontally to one kilogram mass results in an acceleration of one meter per second. This is a departure from traditional English and metric systems, which define force in weight units, caused by the downward vertical acceleration due to gravity. Thus, when the weight of a mass of M kilograms is computed, it is equal to Mg newtons.

First techniques in dimensional analysis convert quantities to MLT or FLT fundamental units. Exponents can then be used to reduce the number of variables and develop equations that establish relationships between these quantities. More advanced techniques which incorporate the Buckingham pi theorem[13] are used to reduce the number of variables to fundamental quantities and predict full-size prototype behavior from model similitude, the study of scale prototype models in hydraulic research.

1-7   STUDY QUESTIONS AND PROBLEMS

1. Define the study of fluid mechanics.
2. What is a fluid?
3. What are the differences between liquids and gases?
4. What is meant by a coherent system of units?
5. List the three classes of SI units.
6. Compute the answers to the following problems, using powers of 10.
    a. $(21.7 \times 10^6) \times (13.4 \times 10^{-2})$
    b. $(16.5 \times 10^4) \times (8.6 \times 10^8)$
    c. $(12.4 \times 10^3) \div (7.2 \times 10^{-4})$
    d. $(4.6 \times 10^{12}) \div (9.7 \times 10^6)$
    e. $(26.2 \times 10^3) \div (1.8 \times 10^{-14})$
7. What makes a digit significant?
8. Round the following numbers to five, four, and three significant digits.
    a. 654 321   d. 213 782
    b. 176 542   e. 479 264
    c. 123 456
9. Describe the difference between the mass of a substance and its weight. What is the major difference in the definitions of the newton force in SI units, and the force unit in other systems, such as the pound in the English system?

---

[13]This theorem is the contribution of E. Buckingham, published under the title, "On Physically Similar Systems," *Physics Review*, Vol. 4 (1914), pp. 354–376.

10. Determine the weight in newtons of three substances whose masses are (a) 5 slugs, (b) 10 kilograms, and (c) 15 grams, respectively.

11. Develop *MLT* and *FLT* expressions for mass flow rate $\dot{M}$, weight flow rate $\dot{w}$, volume flow rate Q, power P, and pressure p.   Develop a table like that in Table 1-7.

12. Derive an *MLT* equation indicating that the power of a hydraulic motor is a function of the pressure and flow rate.

13. Develop an equation for the constant of proportionality k if the period T of a pendulum is proportional to the square root of its length.

14. Develop an equation for the constant of proportionality for a parachute if the velocity of descent is proportional to the square root of its load.

15. If the terminal velocity of a steel ball dropped in a fluid is a function of its diameter, the specific weight of the fluid, and the absolute viscosity of the fluid, develop the equation to solve for this velocity. *Note*: Absolute viscosity μ in *MLT* units is measured as $ML^{-1}T^{-1}$.

2

PROPERTIES OF FLUIDS

2-1 INTRODUCTION

Fluid properties are used to describe their fundamental nature. They may be used, for example, to distinguish fluids from other substances, or, more commonly, to distinguish fluids from each other. Fundamental fluid properties include density, specific weight, specific volume, specific gravity, surface tension, and capillarity. Operational properties that affect the behavior of specific fluids in machines include viscosity, viscosity index, pour point, neutralization number, and antiwear qualities. These have utility in the solution of a wide range of applied problems in fluid mechanics, particularly in hydraulics.

Fluids are divided into two large categories: liquids and gases. Liquids have definite mass and volume, but not definite shape. The shape that the fluid assumes is that of the container which it occupies. If the liquid is poured into an irregular container, it will assume the shape of the inside of the container. If it is poured into a shallow pan or onto the floor, it will assume the shape of the surface and spread in all directions, with the free surface arranged in a plane perpendicular to the force exerted by the gravitational pull of the earth. The plane perpendicular to the gravitational pull of the earth is considered to be level and, for practical purposes, is even with the horizon.

The volume of a liquid is independent of its shape and for most purposes is independent of the pressure exerted on it. Nominal reduction in volume resulting from pressure is about one percent for each $13\,789.5 \times 10^3$ newtons exerted per square meter of surface area. For example, if a 13.79 MN force were exerted on a column of water in a tube 1 m high with an area of 1 m^2, the column could be expected to reduce its height by about 1 cm (Fig. 2-1).

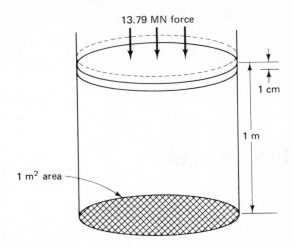

Fig. 2-1   Reduction in volume with pressure

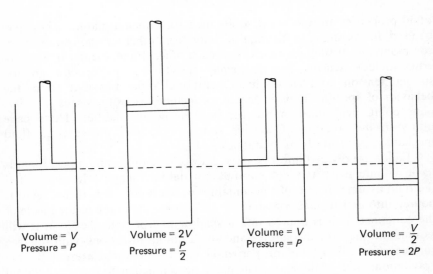

Fig. 2-2   Gas volume and pressure vary inversely

The volume of a liquid is also slightly dependent on temperature and can be expected to change about two percent for each 1° C of temperature change. In closed systems containing only liquid this is critical, and allowances must be made for expansion and contraction as the temperature changes. Automobile cooling systems are an excellent example of this phenomenon and utilize a pressurized filler cap to compensate for liquid expansion by setting a limit on the pressure that increases during operation (usually 50–100 kPa gauge pressure). The cap also allows air to enter the system freely to replace the void left by the fluid as it contracts when cooling.

Gases have definite mass, but no definite shape. They expand or contract to fill the container holding them, and their volume is highly dependent on pressure and temperature. If a container holding one unit of gas under pressure is enlarged to double the available space, the gas contained will expand to occupy the total volume, and the pressure will be reduced proportionally. If the container again assumes its original volume, with no change in temperature, the gas will again occupy the total volume and maintain its original pressure. If the container further reduces the volume occupied by the gas by one-half, with no change in temperature, the gas will still occupy all the space available and the absolute pressure will be doubled (Fig. 2-2).

2-2 DENSITY, SPECIFIC WEIGHT, SPECIFIC VOLUME, AND SPECIFIC GRAVITY

Density ρ (Greek letter rho) is defined as the mass M of a body per unit volume V. In notation

$$\text{Density } \rho = \frac{\text{mass of body } M}{\text{volume of body } V}$$

and

$$\rho = \frac{M}{V} \tag{2.1}$$

An approximation of the density of a substance can be determined experimentally by specifying a given volume and determining its mass with a beam balance. Since the mass is the same on both sides of the balance point, the beam would hang level regardless of changes in gravity. For liquids such as hydraulic oils of different compositions, the volume can be specified in milliliters   (1 cm^3 occupies a volume equal to 1 ml liquid measure) and then balanced against standard weights on a beam balance. The mass of the container should be determined first, and then subtracted from the combined total of the mass of the container and the liquid (Fig. 2-3). Values determined must be in common units. In SI units

$$\rho \text{ Kg/m}^3 = \frac{M \text{ kg}}{V \text{ m}^3}$$

Several empirical standardized test methods have been approved by the American Society for Testing and Materials to determine experimentally the density of hydraulic fluids. Among these are included Test

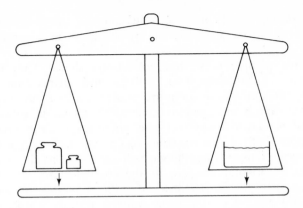

Fig. 2-3 Determining mass with beam balance

Method D 941, in which the density is calculated from the weight of a specified volume of the fluid, and Test Method D 1298, in which the density is read directly from a hydrometer lowered into the sample. Correction factors are applied to the results of these test methods to account for variance in temperature and the buoyancy of air.

The volume of most fluids varies with temperature. As the temperature of a liquid increases, the volume also increases and subsequently the density decreases. To obtain accurate results, the computed values of density for liquids must be corrected to some standard to account for temperature, usually to 15°C. Correction factors for many liquids may be obtained from standard petroleum oil tables.

The specific weight of a substance γ (Greek letter gamma), often referred to as weight density, is defined as the weight ($w = Mg$) of a substance per unit volume V. In notation

$$\text{Specific weight } \gamma = \frac{\text{weight of body } Mg}{\text{volume of body } V}$$

and

$$\gamma\,\text{N}/\text{m}^3 = \frac{M\,\text{kg} \cdot g\,\text{m}/\text{s}^2}{V\,\text{m}^3} \qquad \textbf{(2.2)}$$

Specific weight in newtons per cubic meter (N/m³) or pound-force per cubic ft (lbf/ft³—gravitational unit) has application where the primary concern is weight, for example, where a product is purchased by the cubic meter or cubic foot but is transported by kilogram mass or pound-force weight. In this instance it may be desirable to know if a particular vehicle can transport the mass of a given volume of liquid, the freight charges involved, and the transportation cost per cubic meter of cased product.

Specific volume is defined here as the reciprocal of specific weight. Where specific weight γ is expressed in N/m^3 or lbf/ft^3, specific volume $(u = 1/\gamma)$ is expressed in m^3/N or ft^3/lbf. While specific weight has application where the primary concern is weight, for example, the capability of a vehicle to carry a certain load computed from the weight per cubic meter of the substance, specific volume has application where the primary concern is volume. For example, it may be necessary to determine the capability of a vehicle to contain a certain volume of foam insulation, computed from the specific volume in m^3/N or ft^3/lbf of the substance.

Specific gravity (Sg) is defined as the ratio of the specific weight of a given substance to the specific weight of a standard substance. That is,

$$\text{Specific gravity, } Sg = \frac{\gamma \text{ of substance}}{\gamma \text{ of standard}}$$

and

$$Sg = \frac{\gamma}{\gamma \text{ std}} \tag{2.3}$$

The standard for computing the specific gravity of solids and liquids is water, and for gases is air. Both standards have an assigned value of 1 at standard conditions of 4°C and 76 mm of Hg. For convenience, standard conditions can be approximated at room temperature and atmospheric conditions without introducing appreciable error.

Density and specific weight can be related by substituting the value for density ρ into the formula for specific weight. That is,

$$\gamma = \frac{Mg}{V}$$

$$\rho = \frac{M}{V}$$

Substituting ρ for M/V yields

$$\gamma \, N/m^3 = \rho \, kg/m^3 \cdot g \, m/s^2 \tag{2.4}$$

This relationship indicates that the specific weight of a substance equals the density multiplied by the acceleration due to gravity. It is the equivalent of saying that the force (weight) equals the mass multiplied by the acceleration ($F = Ma$) on a unit basis, which is essentially a restatement of Newton's second law of motion.

The relationship between density and specific gravity is defined from

$$\rho = \frac{\gamma}{g}$$

and since

$$Sg = \frac{\gamma}{\gamma\, std}$$

it follows that

$$\rho = \frac{(\gamma\, std)(Sg)}{g} \qquad\qquad (2.5)$$

2-3   SURFACE TENSION

Surface tension has importance when a free surface is exposed and the boundary conditions are numerically small, for example, in the rise of liquids in capillary tubes of less than 3 mm in diameter, the capillary rise of moisture in soils, and in the formation of mists and sprays.

The sum of the attractive forces acting on a molecule within a liquid body is zero, whereas the cohesive net force acting on a molecule at the surface is inward and perpendicular to the surface because of the unequal force exerted by the medium of the adjacent surface.

The unit surface tension in a liquid, σ (Greek letter sigma), is equal to the work necessary to cause high-energy molecules that form a unit of area at the surface to migrate from inside the liquid. It is measured in work units per area of surface. In notation

$$\text{Surface tension } \sigma = \frac{joules}{meter^2}$$

In *FLT* units,

$$\sigma = \frac{(L)(F)}{L^2}$$

and

$$\sigma = FL^{-1}$$

in units of N/m or lbf/ft. This says, in effect, that the surface tension is numerically equal to the contractile force measured on a line of unit length at the surface of the liquid.

The contractile force of surface tension can be demonstrated by floating a clean needle on a blotter raft and observing that it will float unsupported in water after the blotter becomes saturated and sinks.

Another method is to form a soap bubble at the end of a pitot tube like that shown in Fig. 2-4. From the figure, it can be seen that the net force F_p generating the bubble is equal to the pressure times the cross-sectional area of the tube. That is,

$$F_p = (p)\frac{(\pi d^2)}{(4)}$$

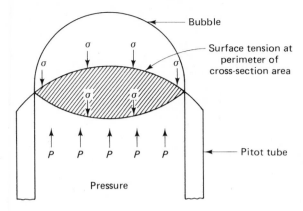

Fig. 2-4 Surface tension forces

The opposite and equal contractile force of surface tension, F_σ, necessary to maintain the bubble around the perimeter and at a tangent to the skin-like membrane is computed from

$$F_\sigma = (\sigma)(\pi d)$$

Equating these two forces and solving for the pressure p, we obtain

$$F_p = F_\sigma$$

$$(p)\left(\frac{\pi d^2}{4}\right) = (\sigma)(\pi d)$$

and

$$p = \frac{4\sigma}{d} \qquad \textbf{(2.6)}$$

which indicates that the pressure within the bubble varies inversely with the diameter. As the size of the bubble increases, the pressure decreases; conversely, as the diameter of the bubble decreases, the pressure increases.

2-4 CAPILLARITY

Capillarity is the tendency for a liquid to become attached to the boundary medium. A drop of water, for example, will spread in all directions on the surface of gasoline or on a flat glass surface. Conversely, mercury will form droplets when touching a flat glass surface.

Liquid molecules thus exhibit an adhesive tendency for dissimilar media either greater or less than the cohesive tendency for molecules of their own kind. This explains the variance in contact angle at the boundary

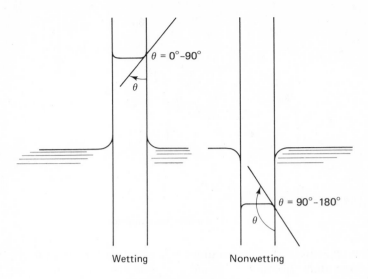

Fig. 2-5   Contact angle

between liquids such as water and mercury and a glass tube of small diameter lowered in them as shown in Fig. 2-5. When the adhesive tendency of the liquid toward the wall surface is greater than the cohesive tendency, the liquid (water) will rise in the tube, and the liquid is said to be *wetting*. When the cohesive tendency of the liquid for itself is greater than the adhesive tendency toward the wall surface, the liquid will fall in the tube, and the liquid is said to be *nonwetting*.

Wetting liquids such as water will have a contact angle upward near the solid boundary between 0 and 90 deg, causing the liquid to rise, whereas nonwetting liquids such as mercury will have a contact angle downward between 90 and 180 deg, causing the liquid to fall.

The weight F_w of the column of fluid supported by the capillary effect is equal to the volume times its specific weight. In notation

$$F_w = (V)(\gamma)$$

and

$$F_w = \left(\frac{\pi d^2}{4} \right)(h)(\gamma)$$

With reference to Fig. 2-6, if the total surface tension σ supporting this column of fluid on a line at an angle θ with the perimeter is set equal to r, the vertical component of this tension x can be computed from

$$x = r \cos \theta$$

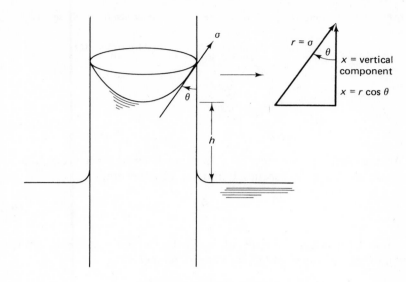

Fig. 2-6 Capillary effect

Substituting σ for r, we obtain

$$x = \sigma \cos \theta$$

and the total vertical component F_x around the perimeter of the boundary tube is

$$F_x = (\pi d)(\sigma \cos \theta)$$

For equilibrium, $F_w = F_x$, and this relationship can be equated and solved for the column height h. That is,

$$F_w = F_x$$

$$\left(\frac{\pi d^2}{4}\right)(h)(\gamma) = (\pi d)(\sigma \cos \theta)$$

and

$$h = \frac{4\sigma \cos \theta}{\gamma d} \qquad (2.7)$$

For wetting fluids and angles between 0 and 90 deg, the value of h will be positive. For nonwetting fluids and angles between 90 and 180 deg, the value of h will be negative.

TABLE 2-1   Specific weight and surface tension of selected liquids at room temperature

Liquid	Specific weight (γ) $N/m^3 \times 10^{-3} =$	Surface tension (σ) $N/m \times 10^2 =$
alcohol	7.69	2.23
castor oil	9.27	3.06
crude oil	8.95	2.72
kerosene	7.85	2.77
mercury	133.37	51.37
water	9.79	7.26

As the wetting tendency of a fluid such as water increases, θ approaches zero, $\cos\theta$ approaches 1, and Eq. (2.7) becomes

$$h = \frac{4\sigma}{\gamma d} \tag{2.8}$$

Table 2-1 lists approximate values for the specific weight and capillary tension of several common liquids at room temperature.

EXAMPLE 2-1

Calculate the rise of water above the surface in a 3-mm (0.12-in.) tube if the wetting angle is assumed to be zero and the specific weight of water[1] at room temperature is 9.79×10^3 N/m^3 (62.31 lbf/ft^3).

SOLUTION
From Eq. (2.8)

$$h = \frac{(4 \times 7.26 \times 10^{-2} \text{ N/m})}{(9.79 \times 10^3 \text{ N/m}^3)(3 \times 10^{-3} \text{ m})}$$

and

$$h = 9.89 \times 10^{-3} \text{ m} \quad \text{or} \quad 9.89 \text{ mm } (0.032 \text{ ft} \quad \text{or} \quad 0.389 \text{ in.})$$

2-5   VISCOSITY

Viscosity is a measure of the internal resistance of a fluid to shear and indicates its relative resistance to flow. Viscosity is related to the internal friction of the fluid itself. Thick fluids flow much more slowly than thin fluids, indicating that their internal friction is higher. Viscosity numbers may be assigned to describe the relative differences in the ability of a fluid

---

[1]When the specific weight of water is given in lbf/ft^3, conversion to N/m^3 is computed from(lbf/ft^3) (N/lbf)(ft^3/m^3)=lbf/ft$^3 \times$157.087 46=N/m^3. For example, the specific weight of water given as 62.31 lbf/ft^3 at room temperature equals 9788.1196 N/m^3.

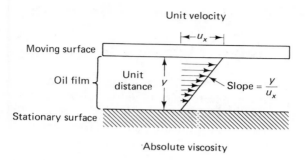

Fig. 2-7 Absolute viscosity

to flow in comparison with other fluids. Higher numbers are assigned to thicker fluids, lower numbers to thinner fluids. Viscosity is important in such applications as hydraulics.

Temperature affects the viscosity of a fluid inversely. That is, as the temperature increases, viscosity decreases. Subsequently, measures of viscosity must be reported with the temperature at which they were determined. Some common standard temperatures at which viscosities are determined are 37.7°C (100°F) and 98.88°C (210°F).

The oil film between a moving part and a stationary part may be thought of as a series of layers of the lubricant separating the two parts. Oil adheres to both surfaces. The velocity of the oil at the stationary surface is 0. The velocity of the oil at the moving surface equals the speed of that surface. Between the two surfaces, the velocity of the oil varies on a gradient (straight line) between 0 at the stationary surface and the speed of the moving part at its surface[2] (Fig. 2-7).

The shear stress between adjacent layers of the oil film as they move is proportional to the viscosity. This shear stress in turn is related to the velocity of the moving surface, and the thickness of the oil film. If the relative velocity of the oil film at the surface of the moving plate supported by the oil film is (u_x), and the thickness of the film is (y), the shear stress (τ) (Greek letter tau) between adjacent layers of the oil film may be derived from

$$\tau = (\mu)\frac{(u_x)}{(y)} \tag{2.9}$$

where μ (Greek Letter mu) is the constant of proportionality. If the moving

---

[2]This is strictly true only for Newtonian fluids where there is a linear relationship between the magnitude of the shear stress and the resulting deformation of the fluid. In non-Newtonian fluids this relationship is nonlinear.

surface has unit area, and the velocity and oil film thickness each are given unit value, then

$$\tau = \mu$$

In this instance, μ is called the absolute or dynamic viscosity, which can now be defined as the force required to move a flat surface of unit area at unit velocity when separated by an oil film of unit thickness.

In the SI system of measurement[3], the units of viscosity are derived from

$$\mu = (\tau)\frac{(y)}{(u_x)}$$

(2.10)

$$\mu = \frac{(N/m^2)(m)}{(m/s)} = N \cdot s/m^2$$

In the English system of measurement, the units of viscosity are derived from

$$\mu = \frac{(lbf/ft^2)(ft)}{(ft/s)} = lbf \cdot s/ft^2$$

and

$$\mu = \frac{(lbf/in.^2)(ft)}{(ft/s)} = lbf \cdot s/in^2 \text{ in reyns}$$

In the traditional metric system of measurement, the units of viscosity are derived from

$$\mu = \frac{(dyne/cm^2)(cm)}{(cm/s)} = dyne \cdot s/cm^2 \text{ in poise}$$

Because the reyn and poise are large units of measure, the microreyn (10^{-6} reyn) and centipoise (10^{-2} poise) are used to simplify calculations. In FLT and MLT units, absolute viscosity has the dimensions

$$\mu = FL^{-2}T$$

and

$$\mu = ML^{-1}T^{-1}$$

[3]The traditional metric unit for absolute viscosity, the poise (P), and the centipoise (cP) are converted to SI units by using $P \times 10^{-1} = N \cdot s/m^2$ and $cP \times 10^{-3} = N \cdot s/m^2$.

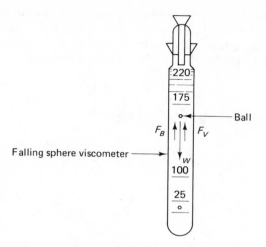

Fig. 2-8   Resolution of forces in falling sphere viscometer apparatus

Absolute viscosity at atmospheric temperature and pressure can be determined in the laboratory by timing the rate at which a falling sphere of specified diameter descends through a fluid. The apparatus includes a falling ball viscometer, hydrometer, and stopclock timer. Referring to Fig. 2-8, when the sphere is falling through the fluid at uniform velocity v, the forces acting on the sphere are gravity (its weight, F_g), the buoyant force of the liquid F_b, and the viscous force of the liquid resisting motion F_v. When the sphere is falling at uniform velocity, the algebraic sum of these forces must be 0. That is,

$$F_g - F_b - F_v = 0 \qquad\qquad (2.11)$$

The weight of the sphere is computed from the density of the material and its volume.

$$F_g = \rho_b g \frac{4}{3}\pi r^3$$

where ρ_b is the density of the ball and r is its radius. The buoyant force F_b of the fluid action on the ball as it descends is computed from

$$F_b = \rho_f g \frac{4}{3}\pi r^3$$

where ρ_f is the density of the fluid. The viscous force F_v is computed from Stokes formula

$$F_v = 6\pi\mu r v$$

where μ is the absolute viscosity of the fluid, r is the radius of the ball, and v is its uniform velocity, computed by dividing the calibrated distance s through which the sphere falls by the elapsed time t.

Solving Eq. (2.11) for the absolute viscosity, we have

$$F_g - F_b - F_v = 0$$

$$(\rho_b g \tfrac{4}{3}\pi r^3) - (\rho_f g \tfrac{4}{3}\pi r^3) - 6\pi\mu rv = 0$$

$$\mu = \frac{\tfrac{4}{3}\pi r^3 g(\rho_b - \rho_f)}{6\pi rv}$$

and

$$\mu = \frac{2}{9}r^2 g\frac{(\rho_b - \rho_f)}{v} \qquad\qquad \textbf{(2.12)}$$

If the Sg rather than the density of the fluid and steel ball is given, Eq. (2.5) is substituted into Eq. (2.12) and thus becomes

$$\mu = \frac{2}{9}r^2 \gamma_{std}\frac{(Sg_b - Sg_f)}{v} \qquad\qquad \textbf{(2.13)}$$

EXAMPLE 2-2

Compute the absolute viscosity of a fluid with Sg of 0.97 through which a 1.5-mm (0.06-in.) ball with a Sg of 7.8 falls at a constant velocity for 75 mm (2.95 in.) in 1 sec.

SOLUTION

Since the specific gravities of the steel ball and fluid are given, viscosity is computed by using Eq. (2.13).

$$\mu = \frac{2}{9}r^2 \gamma_{std}\frac{(Sg_b - Sg_f)}{v}$$

The velocity of descent is computed by dividing the distance through which the ball falls by the elapsed time.

$$v = \frac{(75 \times 10^{-3}\ m)}{1\ s} = 75 \times 10^{-3}\ m/s$$

In SI units

$$\mu = \frac{2}{9}(1.5 \times 10^{-3}\ m)^2(9788.12\ N/m^3)\frac{(7.8 - 0.97)}{(75 \times 10^{-3}\ m/s)}$$

and

$$\mu = 4.46 \times 10^{-1} \, \text{N} \cdot \text{S}/\text{m}^2$$

In traditional metric units, γ_{std} is converted from lbf/ft^3 to dyne/cm^3 by using

$$\text{lbf}/\text{ft}^3 \times \text{N}/\text{lbf} \times \text{dyne}/\text{N} \times \text{ft}^3/\text{cm}^3 = \text{dyne}/\text{cm}^3$$

Substituting gives

$$\gamma_{std} = (62.31 \, \text{lbf}/\text{ft}^3)(4.448 \, 222 \, \text{N}/\text{lbf})$$
$$\times (10^5 \, \text{dyne}/\text{N})(3.53146 \times 10^{-5} \, \text{ft}/\text{cm}^3)$$

and

$$\gamma_{std} = 978.81 \, \text{dyne}/\text{cm}^3$$

Computing the viscosity, we obtain

$$\mu = \frac{2}{9}(1.5 \times 10^{-1} \, \text{cm})^2 (978.81 \, \text{dyne}/\text{cm}^3) \frac{(7.8 - 0.97)}{(75 \times 10^{-1} \, \text{cm}/\text{s})}$$

and

$$\mu = 4.46 \, \text{dyne} \cdot \text{s}/\text{cm}^2 \text{ or poise}$$

Notice that the SI unit for viscosity $(\text{N} \cdot \text{s}/\text{m}^2)$ is 10 times larger than the traditional metric unit, the poise.

In English units

$$\mu = \frac{2}{9}(1.5 \times 32.81 \times 10^{-4} \, \text{ft})^2 (62.31 \, \text{lbf}/\text{ft}^3) \frac{(7.8 - 0.97)}{(75 \times 32.81 \times 10^{-4} \, \text{ft}/\text{s})}$$

and

$$\mu = 9.3 \times 10^{-3} \, \text{lbf} \cdot \text{s}/\text{ft}^2$$

or

$$\mu = (9.3 \times 10^{-3} \, \text{lbf} \cdot \text{s}/\text{ft}^2)(\tfrac{1}{144} \, \text{ft}^2/\text{in}^2)$$
$$= 6.5 \times 10^{-5} \, \text{lbf} \cdot \text{s}/\text{in}^2 \text{ or reyns}$$

Appendix B consists of graphs of the viscosities of several gases and liquids with respect to temperature.

Another direct method used to determine the absolute viscosity of fluids adopted by the American Society for Testing and Materials (ASTM D 88) measures the resistance of fluid to flow by timing a sample of 60 ml

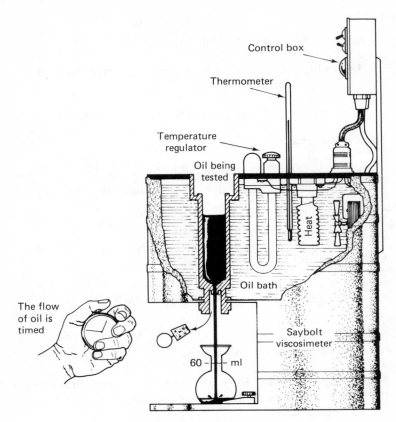

Fig. 2-9   Saybolt viscometer

under a constant head through a standard orifice at the constant temperature of 37.7°C (100°F) or 98.88°C (210°F) (Fig. 2-9). The elapsed time in seconds t is the Saybolt Seconds Universal (SSU) viscosity for the fluid at the given temperature. For thicker fluids, the same test is repeated with a larger orifice to derive the Saybolt Seconds Furol (SSF) viscosity. Furol viscosity time values are approximately 1/10 those obtained by using the Saybolt orifice. The term Furol is a contraction for the words "fuel and road oils."

2-6   KINEMATIC VISCOSITY

Calculations in hydraulics frequently require the use of kinematic viscosity ν (Greek letter nu), rather than the absolute viscosity. Kinematic viscosity is obtained by dividing the absolute viscosity of the fluid by its mass density. That is,

$$\text{Kinematic viscosity}\, \nu = \frac{\text{absolute viscosity}\, \mu}{\text{mass density}\, \rho}$$

and

$$\nu = \frac{\mu}{\rho} \tag{2.14}$$

In *FLT* and *MLT* units, kinematic viscosity has the dimensions

$$\nu = \frac{FL^{-2}T}{FL^{-4}T^2} = L^2 T^{-1}$$

In SI units[4] kinematic viscosity has the units

$$\nu = \frac{N \cdot s/m^2}{N \cdot s^2/m^4} = m^2/s$$

In English units, kinematic viscosity has the units

$$\nu = \frac{lbf \cdot s/ft^2}{lbf \cdot s^2 ft^4} = ft^2/s \ (\text{no special name}) \text{ or } in^2/s \text{ in Newts}$$

In traditional metric units, kinematic viscosity has the units

$$\nu = \frac{dyn \cdot s/cm^2}{dyn \cdot s^2/cm^4} = cm^2/s, \text{ called stokes (St)}$$

In terms

$$\text{Stokes}\, \nu = \frac{\text{poise}\, \mu}{\text{mass density}\, \rho}$$

and since

$$\text{Stokes} \times 0.01 = \text{centistokes}$$
$$\text{Centistokes} = \frac{\text{centipoise}}{\text{mass density}}$$

In the traditional metric system it is common practice in converting from absolute viscosity in centipoise (cP) to kinematic viscosity in centistokes (cSt) to divide by the specific gravity rather than the mass density. If the

---

[4]The traditional metric unit for kinematic viscosity, the centistoke, is converted to SI units by using ν (cSt)$\times 10^{-6} = \nu$ (m^2/s).

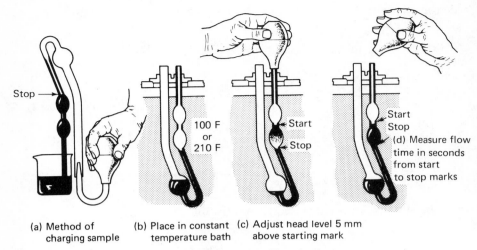

(a) Method of   (b) Place in constant   (c) Adjust head level 5 mm
    charging sample   temperature bath   above starting mark

Fig. 2-10   Kinematic viscometer

mass density of a substance is expressed in gm/cm³, then the numerical values of the specific gravity and the density are the same, since 1 cm³ of water has a density of 1 gm.

The conversion formula changing kinematic viscosity in centistokes to kinematic viscosity in Newts[5,6] is

$$\nu(\text{cSt})(15.52 \times 10^{-4}) = \nu(\text{Newts})$$

A direct method used to determine quickly the kinematic viscosity of fluids in cSt and the absolute viscosity in cP, adopted by the American Society for Testing and Materials (ASTM D 445-65), measures the time required for a fixed amount of an oil to flow through a calibrated capillary instrument using gravity flow at constant temperature (Fig. 2-10). The time is measured in seconds and then multiplied by the calibration constant for the viscometer to obtain the kinematic viscosity of the oil sample in cSt. The absolute viscosity is derived by dividing it by the density.

The conversion from kinematic viscosity in cSt to the equivalent viscosity in SSU is taken or derived from ASTM Procedure D-2161. Basic conversion values given in the procedure are computed for 100°F. For temperatures other than 100°F, kinematic viscosities are converted to SSU viscosities by using temperature correction factors.

When SSU viscosity values are known, values for kinematic viscosity in cSt and Newts may be computed by using two formulas. For elapsed

---

[5]Kinematic viscosity in Newts is converted to kinematic viscosity in centistokes by using ν (cSt) $= \nu$ (Newts) $\times 6 \cdot 4 \times 10^2$.

[6]Conversion of kinematic viscosity in English units to SI units is made by using ν (ft²/s $\times 9.29 \times 10^{-2} = \nu$ (m²/s) and ν (Newts) in²/s $\times 6.452 \times 10^{-4} = \nu$ (m²/s).

times t between 32 and 100 sec and for a controlled temperature of 100°F,

$$\nu(\text{cSt}) = 0.226t - \frac{195}{t} \qquad (2.15)$$

$$\nu(\text{Newts}) = 0.000\ 35t - \frac{0.303}{t} \qquad (2.16)$$

For elapsed times greater than 100 sec and for a controlled temperature of 100°F

$$\nu(\text{cSt}) = 0.220t - \frac{135}{t} \qquad (2.17)$$

and

$$\nu(\text{Newts}) = 0.000\ 39t - \frac{0.21}{t} \qquad (2.18)$$

From the conversion formulas, ν is the computed viscosity value in cSt or Newts, and t is the time in Saybolt Seconds Universal.

EXAMPLE 2-3

An oil is observed to have an SSU viscosity of 80 sec at a temperature of 100°F. What is the equivalent in cSt, Newts, and SI units?

SOLUTION

Since the SSU viscosity is less than 100 sec and the temperature is 100°F, Eq. (2.15) is used.

$$\nu(\text{cSt}) = (0.226 \times 80\ s) - \frac{195}{80\ s}$$

and

$$\nu = 15.64 \text{ cSt}$$

Similarly, using Eq. (2.16), we obtain

$$\nu(\text{Newts}) = (0.000\ 35 \times 80\ s) - \frac{0.303}{80\ s}$$

and

$$\nu = 0.024 \text{ Newt}$$

Finally, converting cSt to SI units, using footnote 3, we have

$$\nu = 15.64 \times 10^{-6} = 1.56 \times 10^{-5} \text{ m/s}^2$$

Approximate conversions among viscosities in SI units, cSt, Newts, and SSU at 100°F can also be made by using Fig. 2-11 which illustrates comparisons of the four measures of viscosity.[7]

---

[7]There are a number of other viscosity measures used in America and throughout the world by industry. A convenient conversion table to convert 42 of them to each other has been published by Gardner Laboratory, Inc., Bethesda, Maryland.

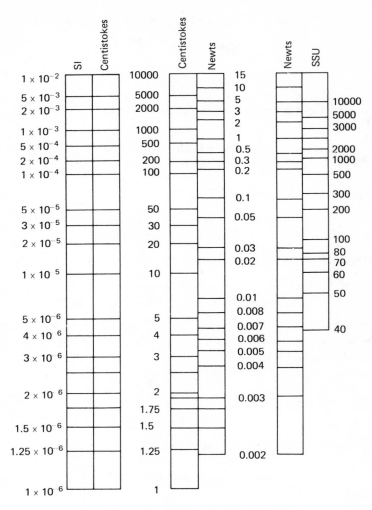

Fig. 2-11   Viscosity conversion chart

2-7   SUMMARY AND RELATED APPLICATIONS

Fluid properties are defined and measured to describe fundamental and operational properties of liquids and gases. "Fundamental" is used here to describe the properties density, specific weight, specific volume, specific gravity, surface tension, and capillarity. Operational property is used to describe those characteristics which affect the usability of the fluid medium in fluid power machinery, for example, the kinematic viscosity, viscosity index, pour point, neutralization number, antiwear properties, and fluid contamination. The behavior of compressible fluids such as air and related properties are highly dependent upon changes in both pressure and temperature, whereas the properties of noncompressible fluids are affected to an appreciable degree by temperature, but only slightly by changes in pressure.

The following related applications have utility in defining fluid properties. Where ASTM standard tests are cited, original source documents should be used as a guide when actual tests are conducted.

1. Capillary effect using a capillary elevation apparatus.
2. ASTM D 941-55, reapproved 1973. Standard Method of Test for Density and Specific Gravity of Liquids by Lipkin Bicapillary Pycnometer.
3. ASTM D 1298-67, reapproved 1972. Standard Method of Test for Density and Specific Gravity of Liquids by Hydrometer Method.
4. Approximation of the density of a substance using the beam balance method.
5. Absolute viscosity of heavy liquids at atmospheric temperature and pressure, the falling sphere method being used.
6. ASTM D 445-74, reapproved 1974. Standard Method of Test for Viscosity of Transparent and Opaque Liquids (Kinematic and Dynamic Viscosities).
7. ASTM D 88-56, reapproved 1968. Standard Method of Test for Saybolt Viscosity.
8. ASTM D 2161-74, reapproved 1974. Standard Method for Conversion of Kinematic Viscosity to Saybolt Universal Viscosity or to Saybolt Fural Viscosity.

2-8   STUDY QUESTIONS AND PROBLEMS

1. What is meant by the term *fluid property*?
2. How is knowledge of fluid properties useful?
3. What is the difference between the specific weight of a substance and its specific gravity?
4. Compute the density of one liter of a liquid that is just balanced by a weight of 15 N.
5. Describe the confusion that exists when the kilogram is defined in terms of both mass and force.
6. Five gallons of a liquid weighs 38 lbf. Compute its specific weight in N/m^3.

7. Water in a container weighs 75 lbf. The same amount of hydraulic fluid weighs 62 lbf. Compute the specific gravity of the fluid.

8. A 55-gal drum of hydraulic fluid weighs 400 lbf. Compute its specific gravity. (*Note*: At room temperature, water weighs approximately 62.3 lbf/ft^3.)

9. Calculate the rise of castor oil above the surface in a 2-mm tube if the wetting angle is assumed to be 45 deg.

10. Using Eq. (2.9), derive the dimensions of absolute viscosity in *FLT* and *MLT* units.

11. Compute the absolute viscosity in SI, traditional metric, and English units of a fluid with a Sg of 0.93, through which a 2-mm (0.079-in.) steel ball with a density of 7.8 falls with a constant velocity of 75 mm (2.95 in.) in 5 sec.

12. If castor oil at 40°C has an absolute viscosity of 2.3×10^{-1} N·s/m^2 (3.34×10^{-5} lbf sec/in.^{2}) and a Sg of 0.86, compute the time it takes for a 1-mm (0.04-in.) steel ball to fall 1 mm (0.04 in.) in a viscosity apparatus like that in Fig. 2-8.

13. Using Eq. (2.14), compute the kinematic viscosity in SI, Newts, and cSt of a fluid with an absolute viscosity of 1 cP.

14. An oil has a SSU viscosity of 90 at a temperature of 100°F. Convert this value to the equivalent viscosity in cSt and Newts.

15. An oil has a SSU viscosity of 300 at a temperature of 100°F. Convert this value to the equivalent viscosity in cSt, Newts, and SI units.

3

FLUID STATICS

3-1 INTRODUCTION

Fluid statics considers the behavior of fluids at rest. Fluids are at rest if they are in equilibrium. That is, the algebraic sum of the external forces acting on the body is zero, and there is little or no relative activity between adjacent layers of the fluid or between the fluid and its confines.

In this condition, the behavior of the fluid is influenced primarily by its elevation, density, temperature, and the effects of gravity.

Fluids at rest exert only normal pressure forces on their confines or on objects placed in the fluid. Because there is little or no relative movement between adjacent layers of the fluid, or between the fluid and adjacent surfaces, there are no shear stresses, and viscosity does not affect the behavior of the fluid or enter into related solutions.

3-2 PRESSURE, AREA, AND FORCE RELATIONSHIPS

Pressure p is defined as force F per area A, where the units must be specified. That is,

$$p = \frac{F}{A} \text{ in N/m}^2 \qquad (3.1)$$

In the SI system, the designated unit of pressure is the pascal (Pa) with units of N/m^2. International Standard ISO 2944-1974(E) defines the unit of pressure as the bar, where

$$1 \text{ bar} = 100 \quad kPa = 14.5 \text{ lbf/in}^2$$

The English unit of pressure is the psi (lbf/in^2). In Europe, the kgf/cm^2

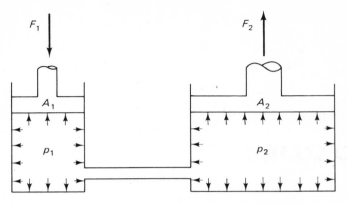

Fig. 3-1 Pressure, force, and area relationships

has been the common unit of pressure.[1] Other designations are used to suit specific applications. Pressure reading gauges indicate the units on the dial. Many incorporate dual scales to display the pressure reading in more than one set of units, although this is illegal in some countries which have converted totally to SI.

Pascal's law states that pressure is transmitted undiminished in all directions throughout a fluid, and that it acts normal to the surfaces of the confines or on any submerged plane. Figure 3-1 illustrates this principle and related pressure, area, and force relationships. Where the pressure in a confined system is p,

$$p = \frac{F_1}{A_1} = \frac{F_2}{A_2} \tag{3.2}$$

EXAMPLE 3-1

A force of 150 N (33.72 lbf) is transmitted from a smaller piston with an area of 25 mm² (0.04 in²) to a larger piston with an area of 100 mm² (0.16 in²). Compute system pressure in pascals, bars, kgf/cm², and lbf/in². Compute the force on the larger piston in newtons and kilograms force.

SOLUTION
With reference to Fig. 3-1, the pressure is computed from

$$p = \frac{F_1}{A_1}$$

$$p = \frac{(150 \text{ N})}{(25 \text{ mm}^2)(10^{-6} \text{ m}^2/\text{mm}^2)} = 6 \times 10^6 \text{ N/m}^2 = 6 \times 10^3 \text{ kPa}$$

[1]The kilogram force (kgf) may still receive limited use in some countries and is related to SI units using (kgf) × 9.806 (m/s²) = N.

Since 1 bar $= 100$ kPa $= 14.5$ lbf/in^2,

$$p = 60 \text{ bars } (870 \text{ lbf/in}^2)$$

In kgf/cm^2,

$$p = \frac{150 \text{ N}}{(9.806 \text{ N/kgf})(25 \text{ mm}^2)(10^{-2} \text{ cm}^2/\text{mm}^2)} = 61.2 \text{ kgf/cm}^2$$

The force on the larger piston, F_2, is computed by solving Eq. (3.2) for F_2.

$$F_2 = \frac{F_1 A_2}{A_1}$$

In newtons

$$F_2 = \frac{(150 \text{ N})(100 \text{ mm})}{(25 \text{ mm})} = 600 \text{ N } (134.9 \text{ lbf})$$

indicating that the multiplication of force is proportional to the ratio of the squares of the piston diameters. Similarly, in kilograms force

$$F_2 = \frac{(150 \text{ N})(100 \text{ mm}^2)}{(9.806 \text{ N/kgf})(25 \text{ mm}^2)} = 61.2 \text{ kgf } (134.9 \text{ lbf})$$

3-3 FLUID-POWER CYLINDERS

Fluid-power cylinders are constructed of a cylinder barrel, piston and rod, end caps, ports, and seals. A typical hydraulic cylinder is shown in Fig. 3-2. The piston provides the effective area against which pressurized fluid is applied and supports the piston end of the rod. The opposite end of the rod is attached to the load which provides the resistance. The cylinder bore, end caps, ports, and seals maintain a fluid-tight system into which fluid energy is piped. Routing of the fluid determines the direction initiated by the piston. Cylinders commonly feature dash pots, snubbers, or decelerators to reduce shock to the mechanism and noise that would be caused by bottoming the piston in either extreme stroke position.

Some fluid-power cylinders are single-acting. That is, they transfer power in one direction only. In a simple application such as that in Fig. 3-3, a single-action piston pump transfers fluid to a single-acting cylinder, causing it to extend. When the cylinder is retracted, the fluid is returned to the reservoir. A ram is a specialized cylinder which uses a large cylinder rod approaching the size of the bore to give maximum support to the load

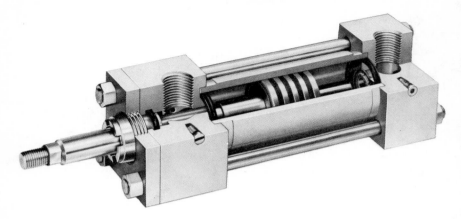

Fig. 3-2 Typical double-acting hydraulic cylinder (*Courtesy of Parker-Hannifin/Cylinder Division*)

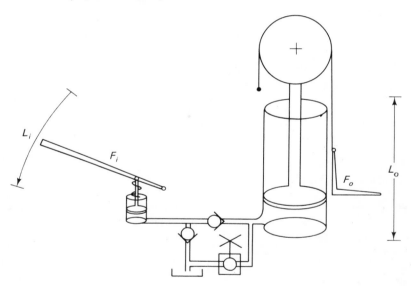

Fig. 3-3 Single-acting piston pump and cylinder

end of the rod. Simple hydraulic jacks, car hoists, and truck bed lifts make extensive use of single-acting rams. Energy required to return the ram to the retracted position must be available in the system.

The output force F required from a hydraulic cylinder and the hydraulic pressure p necessary determine the area A or bore diameter d_b of the cylinder. For a required output force and given pressure,

$$A = \frac{F}{p}$$

and the bore size or diameter

$$d_b = \sqrt{\frac{4F}{\pi p}} \tag{3.3}$$

EXAMPLE 3-2

Compute the cylinder size necessary to move a load resistance of 355 kN (79807 lbf) if the system pressure is 100 bars (1450 lbf/in^2).

SOLUTION
Substituting in Eq. (3.3) in SI units, we obtain

$$d_b = \sqrt{\frac{(4)(35.5 \times 10^4 \text{ N})}{(3.14)(100 \text{ bar} \times 10^5 \text{ N/m}^2)}} = 212.7 \text{ mm} \quad (8.4 \text{ in.})$$

3-4   STRESSES IN CYLINDERS

Cylinders must withstand both circumferencial (hoop) and longitudinal stresses (Fig. 3-4). So long as wall thickness is sufficiently thin—that is, the ratio of the thickness of the wall to the diameter is 0.1 or less—the behavior of the cylinder and related calculations approximate those of thin shell membranes. The internal force F_h that generates the hoop stress tends to split the cylinder in half and can be translated into a pressure p acting against a projected area of width d and length l. That is,

$$F_h = pdl \tag{3.4}$$

The resistance offered by the cylinder to withstand this force S_h equals the product of its tensile strength C_h and the area of the thin wall.

$$S_h = 2C_h t_c l_c$$

To withstand the pressure in the cylinder, the hoop strength of the cylinder S_h must be equal to or greater than the internal radial force resulting from the pressure; that is,

$$S_h \geq F_h$$

and

$$2C_h t_c l_c \geq pdl_c$$

Solving for the tensile strength of the material necessary to withstand the internal force, we obtain

$$C_h = \frac{pd}{2t_c} \tag{3.5}$$

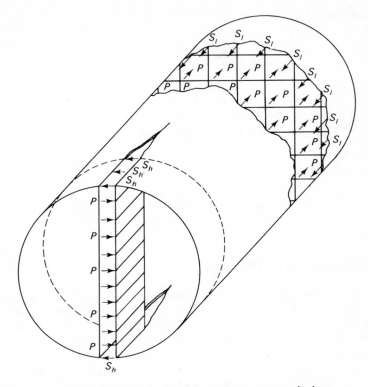

Fig. 3-4   Longitudinal and hoop stresses in a cylinder

The longitudinal force F_l in the cylinder results from pressure acting against the end cap (Fig. 3-4) and equals

$$F_l = \frac{p\pi d^2}{4}$$ **(3.6)**

and the longitudinal strength in the cylinder to withstand this force equals

$$S_l = C_l t_c \pi d$$

where C_l equals the required tensile strength in the longitudinal direction. To withstand the longitudinal stress, the strength of the cylinder must be equal to or greater than the internal longitudinal force resulting from the pressure, and

$$S_l \geqq F_l$$

Solving for the tensile strength of the material to withstand the longitudinal force gives

$$\frac{p\pi d^2}{4} = C_l t_c \pi d$$

and

$$C_l = \frac{pd}{4t_c} \tag{3.7}$$

Comparing the two values of the tensile strength of the material necessary to withstand the internal force resulting from pressure against the inside of the cylinder, we observe that the hoop strength C_h must be twice the longitudinal strength C_l. Subsequently, the tendency for the cylinder to burst is twice that for the cylinder to separate longitudinally, since in actual applications $C_h = C_l$. For this reason, the wall thickness for cylinders under pressure is computed by using Eq. (3.5) and multiplied by a safety factor to ensure system integrity.

EXAMPLE 3-3

Compute the wall thickness of a 100-mm (0.33-ft) cylinder under a pressure of 35 MPa (5075 lbf/in²) if the allowable tensile strength of the material available is 5.17×10^8 N/m² (108×10^5 lbf/ft²) and a safety factor of 2 is applied to ensure system integrity.

SOLUTION

Solving Eq. (3.5) for wall thickness, we obtain

$$t_c = \frac{pd}{2C_h}$$

$$t_c = \frac{(35 \times 10^6 \text{ N/m}^2)(100 \times 10^{-3} \text{ m})}{2(5.17 \times 10^8 \text{ N/m}^2)}$$

and

$$t_c = 3.38 \text{ mm } (0.13 \text{ in.})$$

Multiplying by the safety factor, we obtain

$$t_c \times 2 = 3.38 \times 2 = 6.76 \text{ mm } (0.27 \text{ in.})$$

3-5   PRESSURE, SPECIFIC WEIGHT AND HEIGHT RELATIONSHIPS

From Eq. (1.3), the force exerted by a body at rest is attributed to the effect of gravity acting on its mass. That is,

$$F = w = Mg \text{ newtons (lbf)}$$

where the acceleration due to gravity is 9.806 m/s² or 32.2 ft/s².

Where the volume of a body is specified, dividing the weight by its volume yields the specific weight [Eq. (2.2)]. That is,

$$\gamma = \frac{Mg}{V} = \frac{w}{V} \text{ N/m}^3 \text{ (lbf/ft}^3)$$

Substituting γV for F in Eq. (3.1), we have

$$p = \frac{\gamma V}{A} \qquad \qquad \textbf{(3.8)}$$

in N/m² (lbf/in²) or other consistent pressure units.

If the volume component of Eq. (3.8) consists of a prism of base area A and height h, pressure becomes

$$p = \frac{\gamma \cancel{A} h}{\cancel{A}}$$

and

$$p = \gamma h \qquad \qquad \textbf{(3.9)}$$

where h is in meters, feet, or other linear units, and is called *pressure head* or *head*; it is illustrated in Fig. 3-5.

If the Sg rather than the γ of the fluid is given,

$$p = \gamma_{std} Sg h \qquad \qquad \textbf{(3.10)}$$

These are used to develop several convenient relationships. In SI units, water at standard conditions[2] has a γ_{std} of 9802 N/m² and

$$p = 9802 \, Sgh \quad \text{N/m}^2 \text{ (pascals)}$$

and

$$h = \frac{1.02 \times 10^{-4} p}{Sg} \text{ m}$$

In English units, water has a γ_{std} of 62.4 lbf/ft³ at room temperature or (62.4 lbf/ft³) ($\frac{1}{144}$ in²/ft²) = 0.433 lbf/in² per ft of height, and

$$p = 0.433 \, Sgh \text{ (lbf/in}^2)$$

Or, if pressure is known,

$$h = \frac{2.31 p}{Sg} \text{ ft}$$

---

[2]Standard of water at 15.7°C (60°F) is approximately 9802 N/m³ (62.4 lbf/ft³).

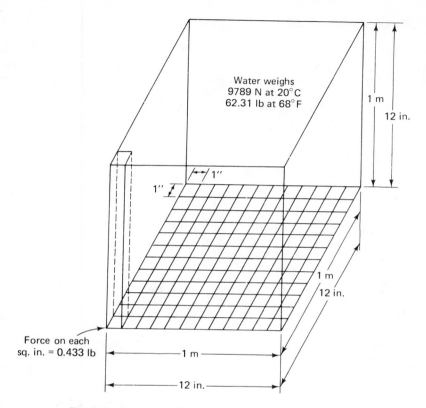

Fig. 3-5   Pressure, specific weight and height relationships

EXAMPLE 3-4

Water in a lake 300 m (984.3 ft) above a hydroelectric plant feeds a turbine. Assuming no losses, compute the pressure in Pa, bars, and lbf/in².

SOLUTION

Substituting in Eq. (3.9), we obtain in SI units

$$p = (9788 \text{ N/m}^3)(300 \text{ m})$$

$$= 29.4 \times 10^5 \text{ N/m}^2 \text{ (pascals)}, \quad \text{or} \quad 29.4 \text{ bars}$$

In English units

$$p = (0.433 \text{ psi/ft})(300 \text{ m} \times 3.2808 \text{ ft/m}) = 426 \text{ lbf/in}^2$$

3-6 PRESSURE MEASURING DEVICES

Both positive and negative pressures are measured from a datum at standard or prevailing atmospheric conditions. Standard atmospheric pressure conditions at sea level are taken as 760 mm (29.92 in.) Hg, which has a Sg of 13.6. Using Eq. (3.10) and $\gamma = 9802$ N/m³ (62.4 lb/ft³), we compute atmospheric pressure in Pa or other convenient units from

$$p = \gamma_{\text{std}} \text{Sg} h$$

and

$$p = (9802 \text{ N/m}^3)(13.6)(0.760 \text{ m}) = 101 \text{ kPa}$$

In English units

$$p = (62.4 \text{ lbf/ft}^3)\left(\frac{1}{144 \text{ in}^2/\text{ft}^2}\right)(13.6)(29.92 \text{ in Hg})\left(\frac{1}{12 \text{ in./ft}}\right)$$

$$= 14.7 \text{ lbf/in}^2$$

Negative pressures below atmospheric are read in inches of H_2O or Hg

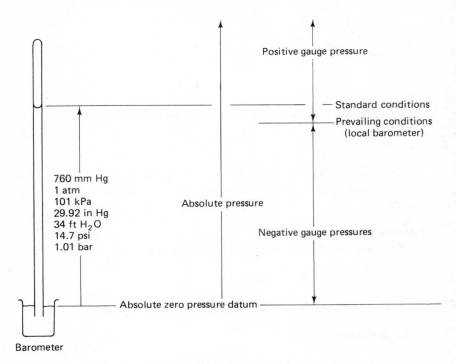

Fig. 3-6 Pressure measurement

from vacuum gauges placed in pump inlet lines or across orifices, although scales in lbf/in^2 and dual scales are also used sometimes for personnel convenience. Pressures below 1 in. of Hg are measured in microns μ, where

$$1\mu = \frac{1}{1,000,000} \ m$$

Figure 3-6 lists common values and units for standard atmospheric conditions. Gauge pressures above atmospheric conditions are read as positive gauge pressures, whereas pressures below atmospheric conditions are read with a negative sign. By convention it is understood that positive gauge pressures do not include local atmospheric conditions, and where absolute pressure is used, it is signified as p abs.

Manometers are simple gauges that measure heights or height differences in tubes filled with liquid. Two of several types available and respective formulas for computing pressure are shown in Fig. 3-7. The

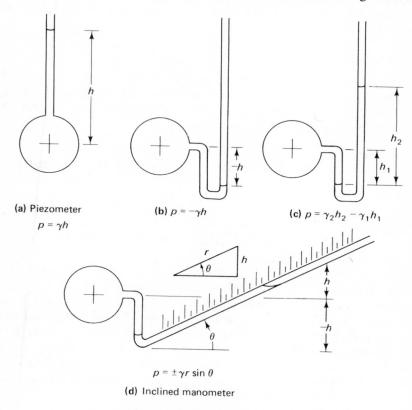

(a) Piezometer
$p = \gamma h$

(b) $p = -\gamma h$

(c) $p = \gamma_2 h_2 - \gamma_1 h_1$

$p = \pm \gamma r \sin \theta$

(d) Inclined manometer

Fig. 3-7 Simple manometers

piezometer shown in Fig. 3-7(a) responds only to positive gauge pressures from liquids with a corresponding rise of liquid in the tube. When equilibrium is reached, the pressure is determined from $p = \gamma h$ in N/m^2 (lbf/in^2) or other units of force per unit area. Because the length of the tube is limited, the piezometer has a narrow pressure range. The inclined manometer shown in Fig. 3-7(d) incorporates a second fluid such as alcohol or mercury and is commonly used for measuring small gas pressures across a calibrated orifice to solve low-velocity gas flow problems. Where the rise $h = r \sin \theta$, if θ conveniently equals 30 deg, two increments of r will equal one increment of h, and pressure is computed from

$$p = \tfrac{1}{2} \gamma r$$

if it is assumed that the reservoir is sufficiently large to maintain the level of the fluid. In practice, after the manometer is leveled and the scale calibrated to zero, the pressure connection is attached to the lower end of the manometer tube, and the change in liquid level is read from the inclined scale.

EXAMPLE 3-5

An inclined manometer with an angle of 30 deg is used to measure the pressure drop across an orifice. If the instrument contains alcohol with a Sg of 0.78 and moves 6 cm (0.2 ft) on the inclined scale, compute the pressure drop in Pa (lbf/in^2).

SOLUTION

Since the Sg rather than γ is given, the pressure drop is computed from

$$p = \tfrac{1}{2} \gamma_{std} Sg r$$

and

$$p = \tfrac{1}{2}(9802 \ N/m^3)(0.78)(6 \times 10^{-2} \ m) = 229 \ Pa \ (0.033 \ lbf/in^2)$$

Figure 3-8 illustrates the application of a differential manometer across an orifice in a pipe. Differential manometers measure pressure differences between two points rather than the gauge pressure, which uses atmospheric conditions as a datum. Valving and fittings permit isolation and removal of the gauge from the system. Standard orifice plates supplied with the differential manometer conveniently scale differential pressures to the calibrated scale on the face of the instrument. By enlarging the surface area at the high-pressure end, the response and sensitivity of the manometer can be increased, since increases in area result in corresponding increases in the output force and resultant liquid height in the low-pressure

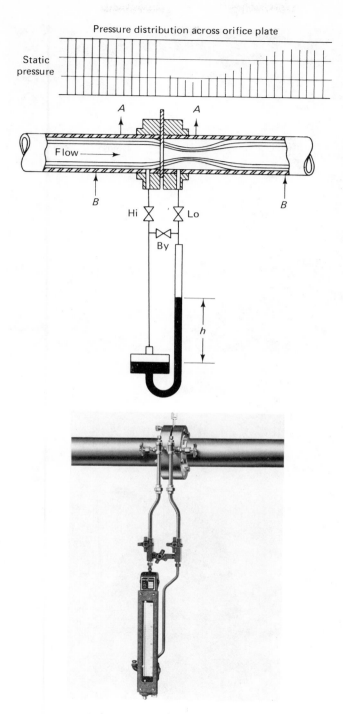

Pressure distribution across orifice plate

Static pressure

Flow

Hi   Lo

By

h

Fig. 3-8 Differential manometer application

Fig. 3-9 Bourdon tube pressure gauge

leg. Doubling the diameter of the pressure-sensitive surface, for example, would quadruple the area and resultant force applied to the liquid surface, resulting in a corresponding displacement of fluid to the smaller leg of the manometer.

Bourdon gauges such as that shown in Fig. 3-9 measure both positive and negative pressures from local atmospheric conditions. They are constructed from a flattened bent tube attached to a movable pointer. Pressures greater than local atmospheric conditions cause the tube to straighten slightly, with the end moving linearly, and the pointer to indicate positive pressures. Vacuum pressures less than local atmospheric conditions will cause the tube to adopt a closer bend and the pointer to indicate negative pressures. Positive pressures are scaled in Pa bars, lbf/in^3, Kg/cm^2, or m (ft) of head, whereas negative pressures are usually scaled in mm (inches) of Hg or H_2O. To insure accuracy, bourdon gauges are sized to systems such that they register normal expected pressures at approximately half scale. Although the cross section of most bourdon tubes is formed over a mandrel, high-quality gauge tubes are often drilled, flattened, and then bent to improve accuracy and gauge integrity.

Most gauge movements are adjustable and calibrated against either an instrument quality bourdon gauge or a dead weight tester, such as that shown in Fig. 3-10. With the dead weight tester connected to the pressure gauge to be calibrated, known hydrostatic forces are applied to the specified cross-section of piston, generating accurate and reliable pressures in the fluid through a wide range of values. Typically, bourdon gauges are calibrated for highest accuracy only within the expected range of operation.

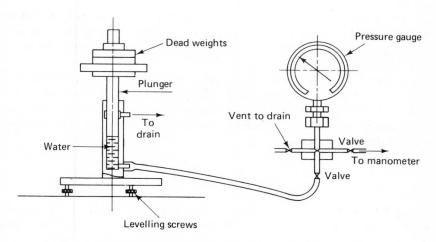

Fig. 3-10   Dead weight pressure gauge tester

3-7   FORCES ON SUBMERGED PLANES

The pressure generated by a fluid with a free surface varies directly as its depth and the direction of pressure is perpendicular to the restraining wall. Such would be the case if the container were a tank, reservoir, lake, or ocean. The magnitude of the pressure p equals γh, if the density of the fluid is given, or $\gamma_{std}Sg\,h$, if the specific gravity is given.

When a plane surface is submerged in a liquid, pressure also acts against its upper and lower surfaces with a compressive tendency even though in a liquid with a free surface the direction of pressure and resultant force in the liquid is upward and toward the surface for positive values of p.

For thin horizontal planes of uniform thickness and material (Fig. 3-11) of surface area A, pressure acts uniformly on the entire surface of the plane. Where $p = \gamma h$ and is constant, the total force (F_h) on one face of the plane equals

$$F_h = \gamma h A = pA \tag{3.11}$$

The resultant point of application of the force is at the average depth $\bar{h}$ at the centroid or center of gravity (c.g.), which is the place about which the gravity forces acting on the plane are equally distributed in all directions. If the horizontal plane is sufficiently thin and rectangular, the centroid is the center of area (c.a.) as well, and may be found by dividing the dimensions of length and width by 2; the resultant force also equals

$$F_h = \gamma \bar{h} A \tag{3.12}$$

For horizontal planes of uniform thickness but varying shapes, the center of area may be found most conveniently by first graphing the area into small equal squares, and then dividing the plane along its length and width into equal areas.

Figure 3-12 illustrates a submerged rectangular plane set at an angle θ with the free surface of the liquid. The pressure in the liquid along the surface of the plane varies with depth h, and since the liquid is at rest, the force due to pressure is exerted perpendicular to the plane. The pressure head at the centroid equals

$$\bar{h} = \frac{h_t + h_b}{2} \tag{3.13}$$

where h_t = depth of the top edge of the plane, h_b = depth of the bottom edge of the plane, and $\bar{h}$ = depth of the c.g. below the liquid surface. The

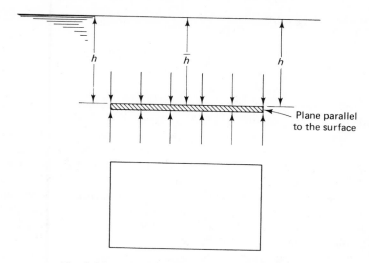

Fig. 3-11 Horizontal plane set below the surface

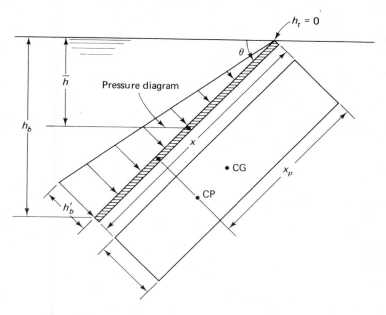

Fig. 3-12 Submerged plane set at an angle with the surface

total force acting on the plane equals

$$F_h = \gamma A \frac{(h_t + h_b)}{2} \tag{3.14}$$

The center of pressure c.p., that is, the point of application of the resultant pressure force, however, is not at the centroid or center of gravity of the plane as was the case for the horizontal plane, but, rather, is below the center of gravity of the inclined plane. The triangular wedge above the surface of the plane represents the distribution of pressure forces exerted on the plane perpendicular to its surface, and it is through the centroid of this triangular projection that the resultant pressure force acts perpendicular to the inclined plane. If the plane is symmetrical, then the center of pressure lies on a line halfway across and running the length of the inclined plane. The centroid of the wedge lies at the intersection of three imaginary lines drawn from the enclosed angles to the midpoints of the opposite sides, and the distance to the centroid of the triangle from the base taken along the inclined plane is one-third its length. Thus, for an inclined plane with the top edge in contact with the surface, the c.p. is always one-third of its length measured from the bottom.

EXAMPLE 3-6

A rectangular plane 0.5 m (1.64 ft) wide and 3 m (9.84 ft) long is submersed to the top edge at an angle of $\theta = 20$ deg with the surface. Compute the vertical distance h to the c.g., the total force on the plane, and the distance to the center of pressure x_p. Refer to Fig 3-13.

SOLUTION

The pressure head at the top of the plate $h_t = 0$, and the pressure head at the bottom edge h_b equals

$$h_b = 3 \sin \theta$$

and

$$h_b = 1.026 \text{ m}$$

The centroid (c.g.) of the plate equals $\frac{3 \text{ m}}{2} = 1.5$ m, and the pressure at the centroid $\bar{h}$ equals

$$\bar{h} = \left(\frac{h_t + h_b}{2} \right)$$

and

$$\bar{h}=0.513 \text{ m}(1.64 \text{ ft})$$

The total resultant force acting against the plate is

$$F=\gamma \bar{h} A$$

and

$$F=(9802 \text{ N/m}^3)(0.513 \text{ m})(0.5\times3.0 \text{ m}^2)=7543 \text{ N } (1696 \text{ lbf})$$

Finally, the center of pressure (c.p.) is computed as the distance one-third the length of the plane measured from the bottom, and from Fig. 3-13:

$$x_p=(x)-\frac{1}{3}(x)=x\left(1-\frac{1}{3}\right)=\frac{2}{3}(x)=\frac{2}{3}(3 \text{ m})=2 \text{ m } (6.56 \text{ ft})$$

putting the c.p. in the lower third of the plane, midway across its width.

When a plane is submerged and set at an angle, the pressure prism is below the surface with a projected area like that shown in Fig. 3-14. The center of gravity is computed as was explained previously, but the center of pressure is derived by first determining the centroid of the projected trapezoid which describes the distribution of the pressure at right angles to the inclined plane, and then locating the point on the plane where a line through the centroid acts perpendicular to the inclined plane. If the plane is symmetrical, then the c.p. also lies on a line halfway across and running the length of the inclined plane. The centroid of the trapezoidal area

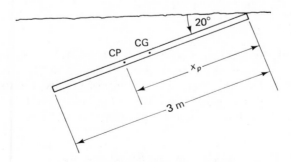

Fig. 3-13 Example 3-6

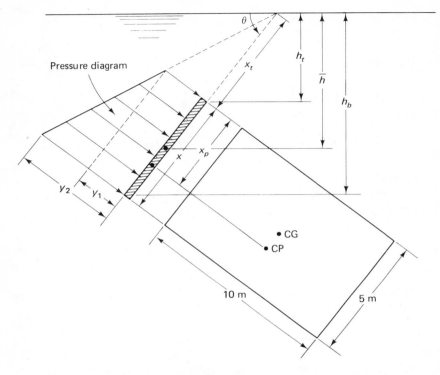

Fig. 3-14   Submerged plane set below the surface at an angle

measured from the top along the inclined plane is given by the formula

$$x_p = x\left[1 - \left(\frac{1}{3}\right)\left(\frac{2y_1 + y_2}{y_1 + y_2}\right)\right] \tag{3.15}$$

An example will illustrate related calculations.

EXAMPLE 3-7

The top edge of a rectangular plane 5 m (16.4 ft) long by 10 m (32.8 ft) wide is submerged at an angle θ of 60 deg, 5 m (16.4 ft) below the surface. Compute the head at the center of gravity, the total force on the surface of the plane, and the center of pressure on the plane surface. Finally, how far is the c.p. below the surface?

SOLUTION
From Fig. 3-14, $h_t = 5$ m and $\theta = 60$ deg, so

$$\bar{h} = 5 \text{ m} + \frac{5}{2}\sin\theta = 5 + 2.5(0.866) = 7.16 \text{ m } (23.5 \text{ ft})$$

The total force on the plane is computed from the head at the centroid, the γ of the liquid, and the area of the plate.

$$F = \gamma \bar{h} A$$
$$F = (9802 \text{ N/m}^3)(7.16 \text{ m})(5 \text{ m} \times 10 \text{ m}) = 3509 \text{ kN (789 lbf)}$$

The center of pressure of the plane is computed from Eq. (3.15). Since by Pascal's law the pressure heads acting about the top and bottom edges of the inclined plane are the same in all directions,

$$y_1 = h_t = 5 \text{ m}$$

and

$$y_2 = h_b = 5 \text{ m} + 5 \sin \theta = 9.33 \text{ m}$$

Substituting Eq. (3.15)

$$x_p = 5\left[1 - \left(\frac{1}{3}\right)\left(\frac{10 + 9.33}{5 + 9.33}\right)\right] = 2.75 \text{ m (9.03 ft)}$$

Finally, computing the vertical distance to the center of pressure

$$h_{c.p.} = 5 \text{ m} + 2.75 \sin \theta = 7.38 \text{ m (24.2 ft)}$$

Notice how this compares to the distance from the center of gravity to the surface, which equals 7.16 m (23.5 ft).

3-8   BUOYANCY AND FLOTATION

Buoyancy is that phenomenon which determines whether an object floats or sinks. When an object floats either above or below the surface in a static fluid, the resultant buoyant force equals the weight of the fluid displaced and acts vertically in the direction of the free surface through the centroid of the displaced volume. This is Archimedes's[3] principle. The resultant buoyant force is nearly independent of the relative pressure that the fluid exerts on the exterior surface of the object because of its depth in the fluid, and equals the difference between the force acting above the object and that acting below the object. Thus if the buoyant force that the fluid exerts is greater or equal to the downward weight of the object, it will float at or in equilibrium below the surface, and if it is not, the object will sink toward the bottom. Figure 3-15 illustrates an example of the application of buoyancy.

[3]Greek philosopher (287–212 B.C.)

Fig. 3-15   An application of buoyancy (*Courtesy of the United States Navy*)

EXAMPLE 3-8

A railroad tie with dimensions 3.5 m × 25 cm × 20 cm floats in water just even with the surface. Determine the buoyant force acting on the underside of the tie.

SOLUTION

The volume of water that the railroad tie displaces equals

$$V = (3.5 \text{ m})(0.25 \text{ m})(0.20 \text{ m}) = 0.175 \text{ m}^3$$

and if the γ of water is taken at 9802 N/m^3, the buoyant force equals

$$F = \gamma V$$

and

$$F = (9802 \text{ N}/\text{m}^3)(0.175 \text{ m}^3) = 1715 \text{ N}$$

In English units, the tie has the approximate dimensions 11.5 ft × 9.8 in × 7.8 in, and the displaced volume and resultant buoyant force would be

$$V = (11.5 \text{ ft})(0.82 \text{ ft})(0.65 \text{ ft}) = 6.1 \text{ ft}^3$$

and

$$F = (62.4 \text{ lbf}/\text{ft}^3)(6.1 \text{ ft}^3) = 381 \text{ lbf}$$

The hydrometer shown in Fig. 3-16 reacts to the buoyant force exerted by a liquid, and depending upon the level at which it floats, indicates the γ and Sg of the liquid. Water with a Sg of 1.00 floats the hydrometer in equilibrium at a mark designated 1.00 on a calibrated scale, making it possible to determine the Sg of liquids of other densities. For practical purposes, the buoyant force of air above the liquid can be ignored.

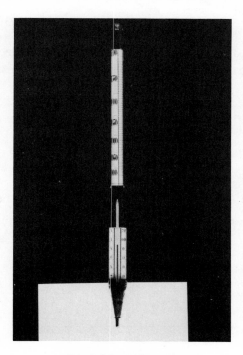

Fig. 3-16 Hydrometer

EXAMPLE 3-9

A wooden piling with a Sg of 0.75 of dimensions 15 m long with a diameter of 30 cm (11.8 in.) is lowered into 12 m (39.4 ft) of water by a crane (Fig. 3-17). Will the pole touch the bottom without additional support or weight?

SOLUTION
The displaced volume of the pole equals

$$V = \left(\frac{\pi d^2}{4} \right)(l)$$

and

$$V = (3.14)(0.09 \text{ m}^2)(0.25)(15 \text{ m}) = 1.06 \text{ m}^3$$

Given a Sg of 0.75, the displaced weight (buoyant force) equals

$$F = \text{Sg}\gamma V$$

and

$$F = (0.75)(9802 \text{ N}/\text{m}^3)(1.06 \text{ m}^3) = 7793 \text{ N}$$

The volume of water with a Sg of 1.00 that the pole displaces equals

$$V_o = \frac{F}{\text{Sg}\gamma}$$

and

$$V_o = \frac{(7793 \text{ N})}{(1.00)(9802 \text{ N}/\text{m}^3)} = 0.79 \text{ m}^3$$

The distance h that the pole can be lowered into the water without

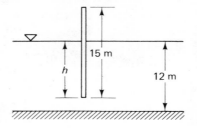

Fig. 3-17   Example 3-9

floating and toppling is

$$h = \frac{4V}{\pi d^2}$$

and

$$h = \frac{(4)(0.79 \text{ m}^3)}{(3.14)(0.09 \text{ m})} = 11.12 \text{ m (36.7 ft)}$$

Thus, the pole would need additional support or weight to remain vertical to the bottom.

3-9 SUMMARY AND APPLICATIONS

Fluids at rest exert forces due to pressure normal to the confines of the fluid and objects immersed in the fluid. Pressure is defined as force per unit area. Common units of pressure are the pascal (N/m^2), bar (100 kPa), kgf/cm^2 (mm of Hg), and the lbf/in^2 (psi).

Pascal's law states that pressure is transmitted undiminished in all directions throughout the fluid, and has extensive application to cylinders and other devices that transmit fluid power from one location to another. Fluid power cylinders receive fluid through ports, which then direct it against the effective area of the piston. Routing determines whether the cylinder extends or retracts. Internal stress tends to both elongate and burst the cylinder barrel, with the tendency for the cylinder to burst being twice that to separate it longitudinally.

Pressure monitoring devices such as manometers and bourdon gauges measure the differences in pressure between a reference pressure such as local atmospheric conditions and a source, or the relative pressure difference between two pressure sources, such as the pressure drop across an orifice plate. Manometers are made more sensitive by inclining the tube through which the fluid is elevated, enlarging the area of the reservoir ends that hold the fluid and against which the pressure is impressed, and varying the Sg of the gauge fluid. The relationship between the force F, γ, the elevation h, and the area A is

$$F = \gamma h A = pA$$

The hydrostatic force on submerged horizontal planes is the product of the fluid density, fluid depth at the centroid, and the area. The point of application of the pressure (c.p.) is also at the centroid (c.g.). If, however, the submerged plane is inclined at an angle, while the expression for the total force remains unchanged because the average depth of the fluid is constant, the point of application of the resultant pressure force is below the center of gravity and halfway across the plane on the line that passes through the centroid of the projected area of the pressure prism.

Buoyancy causes objects to float, the total upward vertical force being equal to the weight of the fluid displaced. If the object is lighter than the weight of the volume of fluid it displaces, it will float, and if not it will sink.

Following are related applications that are useful to develop several common concepts and principles in fluid statics.

1. Balance weights on pistons of different sizes connected together hydraulically to demonstrate Pascal's law and show that the multiplication of forces is inversely proportional to the relative piston areas.

2. Disassemble, clean, inspect, and assemble a hydraulic or pneumatic cylinder. Use a standard overhaul procedure with an appropriate checklist and sign-off.

3. Destructive-test a plexiglass cylinder while monitoring the hydrostatic pressure. Observe whether the failure occurs longitudinally or circumferentially, and compute the stress in the cylinder wall at the time of failure. (*Caution:* The cylinder must be enclosed in a safety apparatus to protect the investigator.)

4. Measure the pressure in a laboratory-constructed system with a U-tube manometer or measure the pressure drop across several standard orifice plates. The design and gauge fluids will determine which formulas are applicable.

5. Service (disassemble, clean, inspect, assemble, and adjust, using instrument tools) and then calibrate several bourdon gauges, using a dead weight tester.

6. Using a laboratory pivot balance apparatus such as that in Fig. 3-18, determine the position of the center of pressure on the rectangular face of the toroid.

7. Construct a simple hydrometer such as that in Fig. 3-19 consisting of a glass tube of known diameter and weight, closed at one end, and into which sand or lead shot is poured. Insert a paper scale in the tube on which the γ and Sg of water are calibrated. Use the constructed hydrometer to compare the γ and Sg of water with those of test fluids such as castor oil, glycerin, and engine oil.

8. Using a laboratory device such as that in Fig. 3-20, verify Archimedes's principle, i.e., that the buoyancy of an object wholly or partly immersed in a fluid experiences an upward thrust equal to the weight of the fluid displaced.

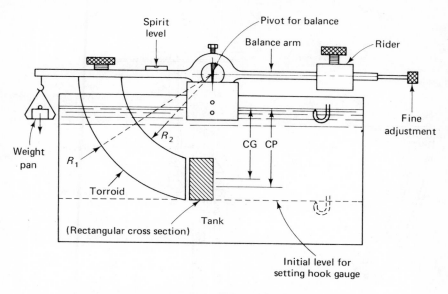

Fig. 3-18 Apparatus for determining the center of pressure

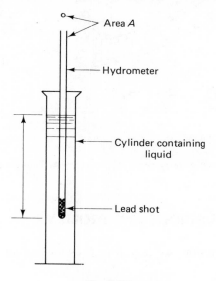

Fig. 3-19

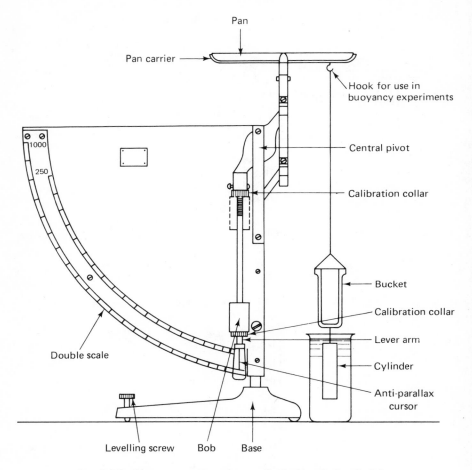

Fig. 3-20 Buoyancy apparatus to verify Archimedes's principle

3-10 STUDY QUESTIONS AND PROBLEMS

1. From the definition, explain why pressure *cannot* flow from a fluid source.

2. Restate Pascal's law.

3. Construct a table of values that lists the force in pounds available from cylinders with diameters of 2, 4, 6, 8, and 10 inches and pressures of 400, 800, 1200, 1600 and 2000 lbf/in². Use a table like that shown

below to record the values

Diameter

2 in.	_____	_____	_____	_____	_____
4 in.	_____	_____	_____	_____	_____
6 in.	_____	_____	_____	_____	_____
8 in.	_____	_____	_____	_____	_____
10 in.	_____	_____	_____	_____	_____
	400	800	1200	1600	2000

Pressure (lbf/in²)

4. Construct a table using the same values as in Problem 3, but this time convert the diameter of the cylinder to centimeters and the pressure to Pa.

5. How does the traditional metric kgf/cm^2 relate to the Pa and the bar?

6. If a hydraulic cylinder has a cylinder rod half the diameter of the bore, what will be the difference in force available if the pressure applied alternately to each end remains constant? (*Clue:* Try two convenient values such as a 4 for the diameter piston and a 2 for the diameter piston rod.)

7. If a hydraulic cylinder has a cylinder rod half the size of the bore, what would occur if pressure were applied to both sides of the piston at the same time? (*Clue:* Make a drawing showing the relative areas of both sides of the piston.)

8. Compute the wall thickness of a 50-mm (1.97-in.) cylinder under a pressure of 100 Mpa (14500 psi) if the material available has an allowable tensile stress of 5.17×10^8 N/m^2 (74.965×10^3 psi). Apply a safety factor of 4 to assure system integrity.

9. Fluid with a Sg of 13 rises 20 cm (7.87 in.) in a simple manometer like that in Fig. 3-7(a). Compute the pressure rise in (a) pascals, (b) kgf/cm^2, and (c) psi.

10. An inclined manometer with an angle of 30 deg and fluid with a Sg of 0.85 is to be used to measure the pressure drop across an orifice at pressures up to 0.05 psi (344.7 Pa). How long must the scale be?

11. A 2-m^3 (70.6-ft^3) cube rests on the horizontal plane 50 m (164.1 ft) below the surface of a lake filled with fresh water. If standard conditions are assumed, what is the crushing force on the top surface of the cube in N and lbf?

12. Figure 3-19 illustrates a square water gate with an area of 1 m² (10.76 ft²) such that the top edge is 10 m (32.8 ft) below the surface of a lake, set at an angle of 60 deg with the surface. The hand lever opens and closes the gate, and the hinge pins about which the gate pivots are set on the center of pressure. What is the total force exerted on the hinge pins, and how far down the gate should the pins be set to allow it to pivot at the center of pressure?

13. A pontoon 75 cm (29.5 in.) in diameter by 3 m (9.84 ft) long that weighs 250 N (56.2 lbf) sinks in fresh water. How much water leaked in?

14. A sealed tank 50 cm (20 in.) in diameter by 3 m (9.84 ft) long that weighs 500 N (112.4 lbf) is to be sunk. How much weight must be added?

15. A man wishes to float a 10-kN (2248-lbf) car in water with balloons that can be inflated to a volume of 1.0 liters (0.035 ft³). If it is assumed that the balloons can be harnessed and can withstand the stress, and that their weight is negligible, how many would be necessary to float the car above and touching the surface?

4

FLOW PHENOMENA

4-1 INTRODUCTION

When a fluid flows, its volume is displaced and changes locations with respect to time. This is true for gases as well as liquids. Important to the understanding are the principles of conservation of mass, from which the continuity equation is developed, kinetic energy, from which several flow equations are generated, and momentum, from which the concepts of work and power are derived. Because the flow of fluids is complex, several of the laws and equations that govern their moving behavior are subject to experiment for validation. As a first step, definitions are useful to structure an understanding of a common body of knowledge about flow phenomena.

A fluid has *boundaries* that separate the control volume, a region of arbitrary dimensions, from the rest of the environment; for example, water is contained by the banks of a river, and air is contained in the pipes of a pneumatic system. While the direction and velocity of flow of the control volume may vary from place to place within the boundaries of the system, at the boundary the velocity of the fluid is zero and there is no flow.

An *ideal fluid* is one that is incompressible and exhibits no viscous shear. In actuality, all fluids exhibit some compressibility and viscous shear. Air, for example, is highly compressible, whereas most liquids are not. An ideal fluid is not to be confused with an ideal gas, which refers to one that follows the gas laws as the volume, pressure, and temperature are varied.

The direction of flow may be *one-dimensional*, and in a straight line; *two-dimensional*, and in parallel planes with the free cross section of flow, uniform in one direction across the stream; or *three-dimensional*, as when the fluid flows in a pipe of changing cross section and the velocity gradient through any cross section of the pipe varies (Fig. 4-1). In one-dimensional flow, particles move tangent to the flow axis, and the fluid within the stream tubes thus formed moves in the same direction. Variations in velocity are usually ignored. Most flow problems consider the flow to be one-dimensional in streams or pipes, and thus consider only a representative stream tube or section from the control volume. The direction of flow is the dimension considered, and thus the cross section of that flow does

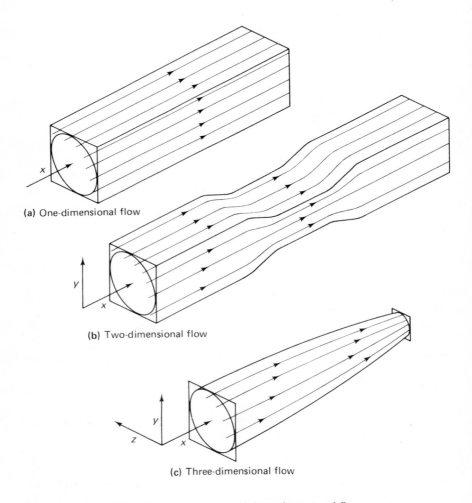

(a) One-dimensional flow

(b) Two-dimensional flow

(c) Three-dimensional flow

Fig. 4-1   One-, two-, and three-dimensional flow

not vary. Two-dimensional flow considers both the direction of flow and one plane within the stream. Flow through the cross section in Fig. 4-1(b), or across an infinitely long rod or airfoil is considered to be two dimensional. Thus, fluid within the stream would be parallel in two planes. Three-dimensional flow occurs when the fluid within the stream tubes in the cross section of flow is nowhere parallel. This occurs when the boundary of the fluid changes cross section, such as when fluid flows through a nozzle, or the pipe changes size or bends.

Steady conditions and flow prevail when successive particles passing a fixed point in the stream have the same velocity over an extended period of time. Thus, changes in pressure, density, flow rate, temperature, etc., would be zero with respect to time. The flow is said to be *uniform* if the direction and magnitude of the velocity do not change between two points distant from each other—for example, in a long pipeline. Flow is *nonuniform* if surging occurs, indicating that the velocity magnitude and/or direction are changing along the line of flow within the control volume.

Streamline flow is indicated when the cross section of flow occurs in smooth laminae. The flow of particles in the streamline is everywhere parallel, and particle flow across the streamline does not occur. *Stream tubes* occur when the fluid is confined in a tubular boundary, such as a pipe, and flow in smooth cylindrical layers. The velocity of the fluid at the conductor wall is zero, whereas the fluid velocity at the center of the conductor and fluid stream is maximum. The velocity gradient across the cross section can be represented by a parabolic curve (Fig. 4-2), indicating that the greatest internal friction and viscous shear in the fluid are near the wall of the conductor.

When *laminar flow* conditions prevail, the viscous nature of the fluid governs its smooth flow, and any inclination toward disorderly turbulent flow is resisted. When fluid flows in a pipe in a disorganized manner, the flow is considered to be *turbulent*, and the smooth streamlines prevalent in laminar flow are disrupted (Fig. 4-3). Agitation within the flowing fluid stream is such that it is difficult to determine the flow path of individual particles, and empirical rather than theoretical solutions to related flow problems are used.

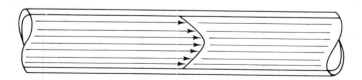

Fig. 4-2 Streamline flow

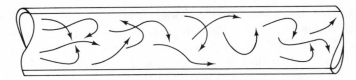

Fig. 4-3   Turbulent flow

4-2   DISPLACEMENT

Displacement in a system refers to a specified volume of fluid and is measured in cubic units of m³ and cm³ (ft³ and in³). It describes the working volume of fluid power components such as pumps, motors, and cylinders during one or more working cycles. Figure 4-4 illustrates a cylinder with a bore d_b, which travels through a stroke s. The area of the bore A_b equals

$$A_b = \frac{\pi d_b^2}{4}$$

and the displacement V_e at the cap end of the cylinder during extension equals

$$V_e = A_b s \qquad\qquad\qquad \textbf{(4.1)}$$

The displacement volume V_r at the head end of the cylinder during retraction equals

$$V_r = s(A_b - A_r) \qquad\qquad\qquad \textbf{(4.2)}$$

where A_r = area of the rod, if it is assumed that the stroke is the same in

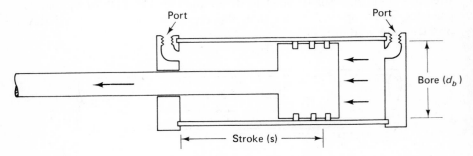

Fig. 4-4   Fluid power cylinder (*Courtesy of Parker-Hannifin/Cylinder Division*)

both directions. If the working cycle consists of both extension and retraction strokes, that is, if the cylinder is double-acting, the displacement through one cycle of operation V_t equals

$$V_t = V_e + V_r \qquad (4.3)$$

and

$$V_t = s(2A_b - A_r) \qquad (4.4)$$

EXAMPLE 4-1

A cylinder with a 75-mm (2.95-in.) bore having a 25-mm (0.98-in.) diameter rod extends and retracts through a 25-cm (9.84-in.) stroke each cycle. Compute the displacement per cycle.

SOLUTION

The areas of the bore and rod are 44.16 cm^2 and 4.91 cm^2, respectively. Substituting in Eq. (4.4), we obtain

$$V_t = 25 \text{ cm}[(2)(44.16 \text{ cm}^2) - (4.91 \text{ cm}^2)]$$

and

$$V_t = 2085 \text{ cm}^3 \ (127 \text{ in}^3)$$

4-3 FLOW RATE

Flow rate Q describes the displacement V per unit time t through a system or some portion of a system. That is,

$$Q = \frac{V}{t} \qquad (4.5)$$

Liquid flow is commonly measured in liters/min and gal/min (gpm). Gas flow is measured in volume measures of standard cubic meters per min or standard cubic feet per min (scfm). The word *standard* refers to the volume that the gas would occupy under one of the accepted standard conditions of atmospheric pressure and temperature, for example, 4°C (39.2°F) and 760 mm Hg (14.7 psi).

Flow rate in cylinders is further subdivided by translating the volume into dimensions of stroke and area.

$$Q = \frac{As}{t} \qquad (4.6)$$

It is noticed, however, that (s/t) in Eq. (4.6) equals the velocity of the cylinder v. Substituting this value in Eq. (4.6), we obtain

$$Q = Av \qquad\qquad\qquad (4.7)$$

where Q, A, and v are in compatible units and each may be solved for in terms of the other two.

EXAMPLE 4-2

What is the capacity of a 30-cm (11.81 in.) diameter circular tile through which water is flowing at 3 m/s (9.84 ft/sec)?

SOLUTION
From Eq. (4.7)

$$Q = (3.14)\frac{(30 \times 10^{-2}\text{m})^2}{4}(3 \text{ m/s})(60 \text{ s/min}) = 12.72 \text{ m}^3/\text{min}$$

Or in English units

$$Q = (12.72 \text{ m}^3/\text{min})(10^3 \text{ l/m}^3)(61.02 \text{ in}^3/\text{l})(1/231 \text{ gal/in}^3) = 3360 \text{ gpm}$$

Fluid flow is measured to monitor the volume flow rate in the system. In hydraulic applications, for example, the flow rate from the pump to an actuator can determine the extent of slippage in components and the volumetric efficiency of the system. Mechanical fluid flow devices in common usage are illustrated in Fig. 4-5 and include flow sight indicators, variable flowmeters, rotameter flowmeters and direct reading flowmeters, which are calibrated to monitor flow in m^3/min, cm^3/min, ft^3/min, gpm, and on other conventional scales.

Variable area flowmeters such as the rotameter create an annular orifice between the sharp edge of the float and the tapered glass tube. A slight pressure drop is present when fluid flows around and past the float. Flow is read as a percentage of full flow between 10 and 100 percent within the limits of the device. One method that can be used to calibrate liquid flowmeters of the variable orifice type at the work site bleeds off and measures the volume of a small percentage of a set flow. For example, 5 or 10 percent of the total flow q is bled off, beginning at full flow, and decreased in increments to about 10 percent, at which time the total flow is diverted and volume flow is measured. Where Q equals the full flow from the system and q equals increments of flow that are diverted and measured,

$$Q = q_1 + q_2 + q_3 + \ldots + q_n \qquad\qquad (4.8)$$

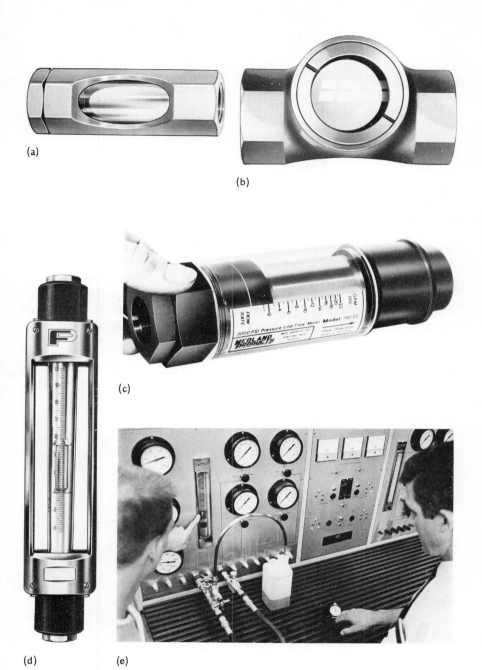

(a)

(b)

(c)

(d)

(e)

Fig. 4-5 Fluid flow monitoring devices: (a–b) Flow sights (*Courtesy of Lube Devices, Inc. and W. E. Anderson, Inc.*); (c) Rotameter (*Courtesy of Fisher and Porter Company*); (d) Direct reading flowmeter (*Courtesy of Hedland Products Division, Racine Federated Inc.*); (e) Flowmeter calibration test set-up (*From Industrial Education, May/June 1971*).

and

$$q_n = \frac{V_n}{t} \text{ liters/min (gpm)} \tag{4.9}$$

Repeated measures are taken at constant temperature and averaged to reduce error. Adding increments starting at 10 percent yields the cumulative flow Q at each succeeding increment. The data and graphic results typically illustrate that the flow characteristic of variable area flowmeters is highly accurate and repeatable, but is rarely linear through the full range of readings.

Because the pressure in pneumatic (gas) systems is relatively constant at the actuator, for example, through the entire stroke of a pneumatic cylinder, air consumption can be computed similarly to fluid consumption in hydraulic systems without appreciable error. Both the velocity and flow rate of the gas are computed at system pressure as though the fluid were incompressible. Allowance is made for the compressible nature of air after consumption at the specified pressure is determined by using Boyle's law for approximate values, or the general gas law, if conversion to standard conditions is specified. Whereas hydraulic systems often measure flow in liquid units (l/m or gpm), pneumatic systems measure flow rate in units of volume (m^3/s or ft^3/sec). An example is used to clarify the calculation.

EXAMPLE 4-3

A single-acting air cylinder with a 5-cm (1.97-in.) bore and a 25-cm (9.84-in.) stroke operates at 8.5 bars (123.3 psi) and cycles at a rate N of 40 cycles per minute (cpm). Compute the air consumption in units of free air.

SOLUTION
The flow volume per minute Q_1 at 8.5 bars (123.3 psi) equals

$$Q_1 = AsN \tag{4.10}$$

$$Q_1 = (3.14)\frac{(5\times10^{-2}\text{ m})^2}{4}(25\times10^{-2}\text{ m})(40 \text{ cyc/min})$$

$$= 19.62\times10^{-3}\text{ m}^3/\text{min}$$

From Boyle's law for isothermal conditions (constant temperature process), free air equals

$$Q_2 = \frac{p_1 Q_1}{p_2} \tag{4.11}$$

and

$$Q_2 = \frac{(8.5 \times 10^5 \ \text{N/m}^2 + 1.01 \times 10^5 \ \text{N/m}^2)(19.62 \times 10^{-3} \ \text{m}^3/\text{min})}{(1.01 \times 10^5 \ \text{N/m}^2)}$$

$$= 0.185 \ \text{m}^3/\text{min (free air)}$$

If temperature correction is desired—for example, if air at the work station is delivered at 25°C and correction to 20°C is necessary—the correction factor is derived from

$$Q_2 = \left(\frac{p_1 Q_1}{p_2}\right)\left(\frac{T_2}{T_1}\right) \tag{4.12}$$

and

$$Q_2 = (0.185 \ \text{m}^3/\text{min})\left(\frac{20+273}{25+273}\right) = 0.182 \ \text{m}^3/\text{min free air at 20°C}$$

In English units

$$Q_1 = (19.62 \times 10^{-3} \ \text{m}^3/\text{min})(61\,024 \ \text{in}^3/\text{m}^3)(1/1728 \ \text{ft}^3/\text{in}^3)$$

$$= 0.69 \ \text{cfm}$$

$$Q_2 = \frac{(8.5 \times 14.5 \ \text{psi} + 14.7 \ \text{ps}i)(0.693 \ \text{ft}^3/\text{min})}{(14.7 \ \text{psi})}$$

$$= 6.5 \ \text{cfm free air}$$

Finally,

$$Q_2 = (6.5 \ \text{cfm})\left(\frac{68+460}{77+460}\right) = 6.41 \ \text{cfm free air at 68°F}$$

Although barometric changes caused by altitude, weather conditions, and relative humidity normally only slightly affect atmospheric conditions (about $2\frac{1}{2}$ percent/1000 ft), they must be considered when the air consumption of components and the necessary delivery of the compressor are computed. Standard air and flow are defined at a temperature of 20°C (68°F), at a pressure of 760 mm of Hg (14.7 psi), and a relative humidity of 36 percent (0.0750 density). This is in agreement with definitions adopted by ASME, although in the gas industries the temperature of *standard air* is usually given as 60°F (15.6°C).[1]

[1]*Compressed Air and Gas Handbook*, 4th ed. New York: Compressed Air and Gas Institute, 1973, p. 10-5.

4-4   CONTINUITY EQUATION

For purposes of making most calculations, the flow in a system is considered to be steady. The flow is steady when the velocity for any period of time is constant. The velocity at successive places in the conduit may change while the flow remains steady, however, for example, where the cross-section area of the fluid conduit is larger or smaller. At places of transition, the fluid velocity changes and the flow is unsteady. To approximately fulfill the conditions of steady flow, changes at a given point with respect to time in pressure, temperature, fluid density, and flow rate would have to be almost zero.

The continuity equation governing flow in a fluid conduit is derived from the principle that during steady flow conditions, the mass of fluid flowing past any section in the conduit is the same. That is,

$$Q = \rho_1 A_1 v_1 = \rho_2 A_2 v_2 = \text{constant} \qquad \textbf{(4.13)}$$

In the case of water and hydraulic fluids, which are nearly incompressible, and for most practical purposes where the density of a fluid between point 1 and point 2 does not change

$$Q = A_1 v_1 = A_2 v_2 = \text{constant} \qquad \textbf{(4.14)}$$

EXAMPLE 4-4

Hydraulic fluid is flowing from a 5-cm (2-in.) diameter pipe to a 10-cm (4-in.) diameter pipe at the rate of 150 liters/min (39.6 gpm). Compute the velocity of the fluid in both pipes.

SOLUTION

With reference to Fig. 4-6, the area of the 5-cm diameter section of pipe is

$$A = \frac{(3.14)(5 \times 10^{-2} \text{ cm})^2}{(4)} = 1.96 \times 10^{-3} \text{ m}^2$$

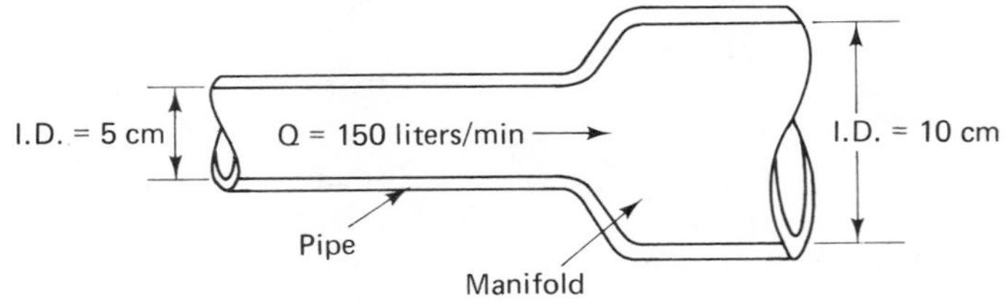

I.D. = 5 cm   Q = 150 liters/min →   I.D. = 10 cm

Pipe

Manifold

Fig. 4-6   Example 4-4

Solving Eq (4.7) for the velocity of the fluid in the 5-cm section, we obtain

$$v_1 = \frac{Q}{A} = \frac{(150 \times 10^{-3} \text{ m/min})(\frac{1}{60} \text{ min/s})}{(1.96 \times 10^{-3} \text{ m}^2)} = 1.28 \text{ m/s } (4.18 \text{ ft/s})$$

The area of the 10-cm diameter section of pipe is four times that of the 5-cm diameter section of pipe, i.e., 7.84×10^{-3} m². Finally, substituting in the continuity equation, Eq. (4.14) and solving for v_2, we obtain

$$v_2 = \frac{A_1 v_1}{A_2} = \frac{(1.96 \times 10^{-3} \text{ m}^2)(1.28 \text{ m/s})}{(7.84 \times 10^{-3} \text{ m}^2)} = 0.32 \text{ m/s } (1.05 \text{ ft/s})$$

EXAMPLE 4-5

The discharge pipe from a pump rated at 8000 liters/min (2113 gpm) emptying reservoir is 15 cm (5.9 in.) in diameter. Compute the velocity of the fluid in the discharge line. Determine the diameter of the suction line if the velocity is not to exceed 2.5 m/s (8.2 ft/s).

SOLUTION

Solving for the velocity of the fluid in the discharge line, we have

$$v_1 = \frac{Q}{A_1} = \frac{(4)(8000 \times 10^{-3} \text{ m}^3/\text{min})(1/60 \text{ min/s})}{(3.14)(15 \times 10^{-2} \text{ m})^2}$$

$$= 7.55 \text{ m/s } (24.8 \text{ ft/s})$$

Finally, solving for the diameter of the suction line, we obtain

$$A_2 = \frac{Q}{v_2}$$

$$d_2 = \sqrt{\frac{4Q}{\pi v_2}} = \sqrt{\frac{(4)(8000 \times 10^{-3} \text{ m}^3/\text{min})(1/60 \text{ min/s})}{(3.14)(2.5 \text{ m/s})}}$$

$$= 26 \text{ cm } (10.2 \text{ in.})$$

Similar values are obtained using flow nomographs, such as those shown in Figs. 4-7 and 4-8, that have been derived using the continuity equation. Knowing two values allows for the solution of the third value by placing a straightedge across two known values and reading the third quantity directly.

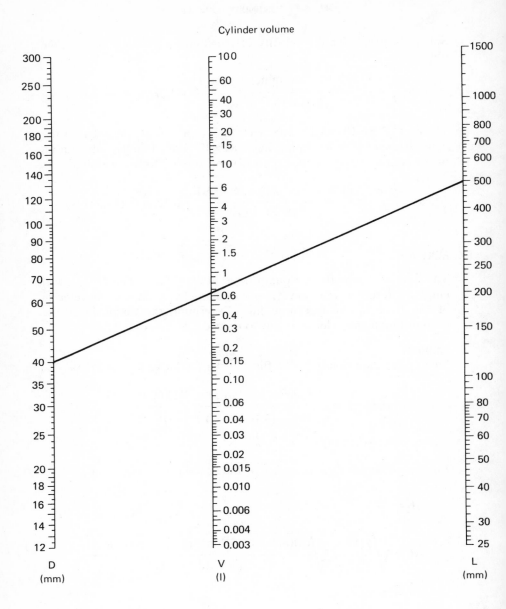

Cylinder volume

Solves the equation $V = KD^2L$, where V is the cylinder volume in litres, D is the cylinder diameter in mm, L is the stroke length in mm, and $K = 7.854 \times 10^{-7}$.

Example: when $D = 40$ mm and $L = 500$ mm, $V = 0.63$ litres

Fig. 4-7   SI units flow nomograph

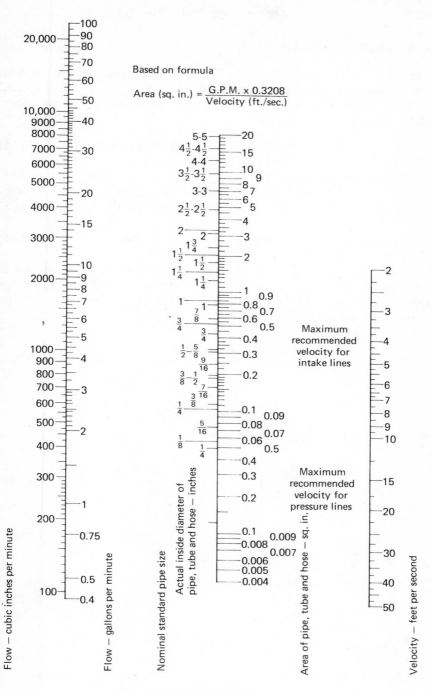

Fig. 4-8 English units flow nomograph

4-5 ENERGY EQUATION

The total energy available from fluid in a system consists of the potential energy due to its initial elevation PE, the flow-work energy FE generated by fluid power components, and the kinetic energy KE due to the movement of the fluid. In notation

$$\text{Total energy} = PE + FE + KE = \text{constant} \qquad \textbf{(4.15)}$$

The potential energy of a fluid can be equated to the force that it exerts toward the center of gravity ($w = Mg$) and the distance h through which it is available to move on command (Fig. 4-9). This is expressed by

$$PE = Mgh = wh \text{ in } \text{m} \cdot \text{N or ft-lbf} \qquad \textbf{(4.16)}$$

The flow-work energy FE is developed or consumed by fluid power components such as turbines, pumps, and motors and incorporates a pressure term p rather than height h as in Eq. (4.16). Where the specific gravity is given, since from Eq. (3.10)

$$p = \gamma_{\text{std}} \text{ Sg } h$$

and

$$h = \frac{p}{\gamma_{\text{std}} \text{Sg}}$$

Substituting for h in Eq. (4.16), we obtain

$$FE = wh = \frac{wp}{\gamma_{\text{std}} \text{Sg}} = \frac{1.02 \times 10^{-4} \, wp}{\text{Sg}} \qquad \textbf{(4.17)}$$

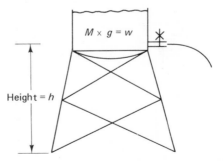

Fig. 4-9 Potential energy as a force available to act through an elevation h

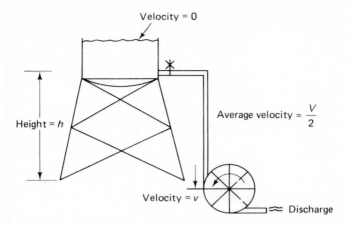

Fig. 4-10 Kinetic energy as the energy of motion

where p is in Pa, i.e., N/m^2. In English units

$$FE = \frac{2.31 \, wp}{Sg} \qquad (4.18)$$

where p is in lbf/in^2.

When fluid flows in a system, potential energy is converted to the energy of motion, kinetic energy (Fig. 4-10). Potential energy and kinetic energy are mutually convertible. As the potential energy of the fluid is dissipated, for example, through a falling action from a higher point to a lower point, its kinetic energy is increased by an equal amount. This is a restatement of the law of conservation of energy, which says, in effect, that energy can be neither created nor destroyed.

If a body of fluid at rest at some height is dropped through a distance h, its final velocity is v m/s. The acceleration due to gravity g is approximately constant. The average velocity of the falling body of fluid is $(v/2)$ m/s, since it has a starting velocity of 0 m/s, and a final velocity of v m/s. The time taken for the body to cover the distance equals the distance h divided by the average velocity $(v/2)$. That is,

$$t = \frac{h}{\dfrac{v}{2}} = \frac{2h}{v} \qquad (4.19)$$

where the time t is in seconds. The constant acceleration due to gravity g

equals the change in velocity divided by the time. That is,

$$g = \frac{v-0}{t} = \frac{v}{t} \qquad (4.20)$$

$$g = \frac{v}{\dfrac{2h}{v}} = \frac{v^2}{2h}$$

and

$$\frac{v^2}{2} = gh \qquad (4.21)$$

From the potential energy formula

$$PE = Mgh$$

where the value $v^2/2$ can be substituted for gh, converting it to the kinetic energy formula. That is,

$$KE = \frac{Mv^2}{2}$$

Substituting w/g for mass M, we obtain

$$KE = \frac{wv^2}{2g} \qquad (4.22)$$

The expression for the total energy in the system now becomes

Total energy $= PE$ (elevation) $+ FE$ (flow-work) $+ KE$ (kinetic)

and

$$\text{Total energy} = (wh) + \left(\frac{1.02 \times 10^{-4}\, wp}{Sg} \right) + \left(\frac{wv^2}{2g} \right) \text{ in m-N} \qquad (4.23)$$

or

$$\text{Total energy} = (wh) + \left(\frac{2.31\, wp}{Sg} \right) + \left(\frac{wv^2}{2g} \right) \text{ in ft-lb} \qquad (4.24)$$

Conservation of energy in fluid power systems is described by Bernoulli's theorem, which says, in effect, that the total energy of the system remains constant (Fig. 4-11). For example, if no work is done or energy dissipated to the surroundings, the energy level of fluid flowing in a streamline through a restriction will remain constant, even though both the pressure and velocity of the fluid change. Bernoulli's theorem equates the

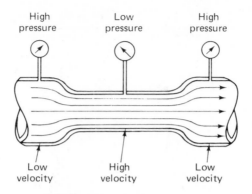

Fig. 4-11   Bernoulli's theorem

energy at any two points in the system. That is,

$$\text{Total energy}_{(\text{point 1})} = \text{total energy}_{(\text{point 2})}$$

and

$$(w_1 h_1) + \left(\frac{1.02 \times 10^{-4} w_1 p_1}{\text{Sg}} \right) + \left(\frac{w_1 v_1^2}{2g} \right)$$

$$= (w_2 h_2) + \left(\frac{1.02 \times 10^{-4} w_2 p_2}{\text{Sg}} \right) + \left(\frac{w_2 v_2^2}{2g} \right)$$

For most calculations, the weight of the fluid w does not change appreciably and will cancel out.[2] And if the potential energy due to elevation is equated to z, the equation becomes

$$z_1 + \left(\frac{1.02 \times 10^{-4} p_1}{\text{Sg}} \right) + \left(\frac{v_1^2}{2g} \right) = z_2 + \left(\frac{1.02 \times 10^{-4} p_2}{\text{Sg}} \right) + \left(\frac{v_2^2}{2g} \right) \quad \textbf{(4.25)}$$

in SI units. In English units

$$z_1 + \left(\frac{2.31 p_1}{\text{Sg}} \right) + \left(\frac{v_1^2}{2g} \right) = z_2 + \left(\frac{2.31 p_2}{\text{Sg}} \right) + \left(\frac{v_2^2}{2g} \right) \quad \textbf{(4.26)}$$

The general and more recognizable form of Eqs. (4.25) and (4.26) replaces

---

[2]From Chapter 3 the γ_{std} for water is taken as 9802 N/m³ and $h = p/\gamma_{\text{std}}$ Sg $= (1.02 \times 10^{-4}$ m³/N$)(p)/(\text{Sg})$.

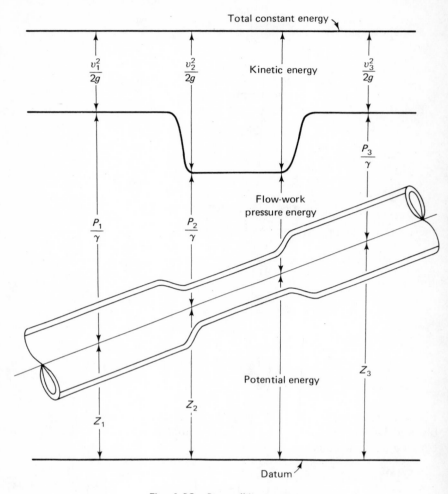

Fig. 4-12 Bernoulli's equation

γ_{std} Sg with γ for fluid and is written as

$$z_1 + \left(\frac{p_1}{\gamma}\right) + \left(\frac{v_1^2}{2g}\right) = z_2 + \left(\frac{p_2}{\gamma}\right) + \left(\frac{v_2^2}{2g}\right) = \text{constant} \qquad \textbf{(4.27)}$$

where each of these terms is in units of length that correspond dimensionally to head. Thus, the term z is the elevation head, the term p/γ is the flow-work or pressure head, and the term $v^2/2g$ is the velocity head. Their relationship is illustrated by Fig. 4-12, which indicates that their sum equals the constant total energy, though they may vary in combination with variation in flow conditions.

Solving problems with Bernoulli's equation requires a systematic approach. In general, the following steps are desirable:

1. Diagram the system, indicating at which points solutions are to be made and the direction of flow.
2. Determine the datum planes and the elevations z_1 and z_2. If these are approximately equal—for example, when fluid flows horizontally in a hydraulic system—they can be cancelled out.
3. Set up Bernoulli's equation with each energy component in the same units, including energy added by pumps in the left member, and energy extracted by turbines in the right member.
4. Flow losses should be added in the right member.
5. If the velocity of the fluid is not known, solve the continuity equation.
6. Finally, equate the total energy at point 1 in the system, including that added and subtracted, to the total energy at point 2 in the system. Typically, the equation is solved for pressure or head, including losses, from point 1 to point 2. For more complete analysis, determine and plot the energy line through the system.

EXAMPLE 4-6

In Fig. 4-13, fluid with a Sg of 0.85 is flowing horizontally at the rate of 2000 liters/min (528.3 gpm) from a pipe at point 1 with an inside diameter of 10 cm(3.9 in.) to one with a 3-cm(1.2-in.) diameter at point 2. The pressure at point 1 is 6.5 MPa (942.5 psi). Assuming that no work is done or energy dissipated from the system, compute the pressure at point 2.

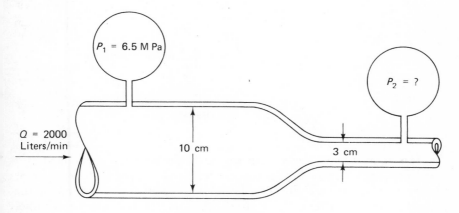

Fig. 4-13   Example 4-6

SOLUTION

Because the flow is horizontal, introducing no change in elevation, the quantities z_1 and z_2 can be cancelled. And since the Sg of the fluid is given, Eq. (4.24) is written

$$\left(\frac{1.02 \times 10^{-4} p_1}{\text{Sg}}\right) + \left(\frac{v_1^2}{2g}\right) = \left(\frac{1.02 \times 10^{-4} p_2}{\text{Sg}}\right) + \left(\frac{v_2^2}{2g}\right)$$

and the direction of flow is from p_1 to p_2. The velocities v_1 and v_2 are not known and must be determined by solving the continuity equation.

$$Q = A_1 v_1$$

and

$$v_1 = \frac{(4)(2000 \times 10^{-3}\ \text{m}^3/\text{min})(1/60\ \text{min/s})}{(3.14)(10 \times 10^{-2}\ \text{m})^2} = 4.25\ \text{m/s}$$

Solving for v_2, we obtain

$$v_2 = \frac{(4)(2000 \times 10^{-3}\ \text{m}^3/\text{min})(1/60\ \text{min/s})}{(3.14)(3 \times 10^{-2}\ \text{m})^2} = 47.18\ \text{m/s}$$

Finally, solving Bernoulli's equation for p_2, we have

$$\frac{(1.02 \times 10^{-4}\ \text{m}^3/\text{N})(6.5 \times 10^6\ \text{N/m}^2)}{(0.85)} + \frac{(4.25\ \text{m/s})^2}{(2)(9.8\ \text{m/s}^2)}$$

$$= \frac{(1.02 \times 10^{-4}\ \text{m}^3/\text{N})(p_2)}{(0.85)} + \frac{(47.18\ \text{m/s})^2}{(2)(9.8\ \text{m/s}^2)}$$

$$(780.00\ \text{m}) + (0.92\ \text{m}) = \frac{(1.02 \times 10^{-4}\ \text{m}^3/\text{N})(p_2)}{(0.85)} + (113.57\ \text{m})$$

and

$$p_2 = 5.56\ \text{MPa (806 psi)}$$

It is noticed that a decrease in the size of the pipe is accompanied by a corresponding increase in velocity, since the flow rate in the system is constant. The pressure drops because the energy transfer through the system remains constant. If the size of the conduit were increased rather than decreased, the velocity would decrease with a corresponding increase in pressure. Finally, the kinetic energy term or velocity head in Bernoulli's equation is seen to increase or decrease as the square of the velocity.

Flow-work energy added to the system by fluid power pumps H_a, energy extracted H_e, and energy lost H_l also must be accounted for when actual rather than ideal flow conditions are computed. These are written in terms corresponding to head or pressure. The complete equation in SI units for Bernoulli's theorem thus becomes

$$z_1 + \left(\frac{1.02 \times 10^{-4} p_1}{Sg} \right) + \left(\frac{v_1^2}{2g} \right) + H_a$$

$$= z_2 + \left(\frac{1.02 \times 10^{-4} p_2}{Sg} \right) + \left(\frac{v_2^2}{2g} \right) + H_e + H_l \qquad \textbf{(4.28)}$$

EXAMPLE 4-7

Fluid flows from a pump at 65 bars (942.5 lbf/in²) at the rate of 8000 liters/minute (2113 gpm) horizontally to a motor operating at 50 bars (725 psi) (Fig. 4-14). Back pressure on the motor discharge port is 25 bars (362.5 psi). The fluid has a Sg of 0.85. Compute the energy extracted from the fluid in terms corresponding to head.

SOLUTION
Flow is horizontal, introducing no change in elevation, and the quantities z_1 and z_2 cancel out of Eq. (4.25). Since the fluid flow rate and pipe size are constant, the velocity head terms are approximately equal

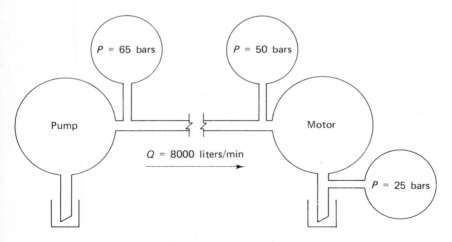

Fig. 4-14 Example 4-7

and also cancel. If the pressure at the inlet to the pump is considered to be 0 bar gauge, the Bernoulli equation for the system then becomes

$$H_a = \left(\frac{1.02 \times 10^{-4} p_2}{Sg} \right) + H_e + H_l$$

Computing the head associated with the flow-work energy added to the system, we obtain

$$H_a = \left(\frac{1.02 \times 10^{-4} p_1}{Sg} \right)$$

and

$$H_a = \frac{(1.02 \times 10^{-4} \text{ m}^3/\text{N})(65 \times 10^5 \text{ N/m}^2)}{(0.85)} = 780 \text{ m } (2559 \text{ ft})$$

Finally, solving Bernoulli's equation for head associated with the energy that can be extracted from the system, we have

$$H_e = H_a - \left(\frac{1.02 \times 10^{-4} p_2}{Sg} \right) - H_l$$

$$H_e = 780 \text{ m} - \frac{(1.02 \times 10^{-4} \text{ m}^3/\text{N})(25 \times 10^5 \text{ N/m}^2)}{(0.85)}$$

$$- \frac{(1.02 \times 10^{-4} \text{ m}^3/\text{N})(15 \times 10^5 \text{ N/m}^2)}{(0.85)}$$

and

$$H_e = 780 \text{ m} - 300 \text{ m} - 180 \text{ m} = 300 \text{ m } (984 \text{ ft})$$

The specific weight of gases is much lighter than that of liquids, so the aerodynamic form of Bernoulli's equation can be simplified. If Eq. (4.27) is multiplied through by the specific weight of the fluid γ, it becomes

$$\gamma z + p + \frac{\gamma v^2}{2g} = \text{constant}$$

It is noticed that γz is very small and can be deleted without introducing appreciable error, and that $\gamma/g = \rho$, the mass density of the gas. Thus, the

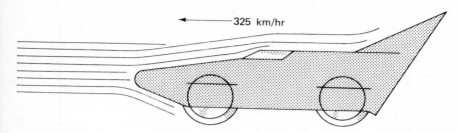

Fig. 4-15   Example 4-8

aerodynamic form of Bernoulli's equation reduces to

$$p_1 + \tfrac{1}{2}\rho v_1^2 = p_2 + \tfrac{1}{2}\rho v_2^2 = \text{constant} \qquad (4.29)$$

where each of these terms is in units of pressure. The p_1 and p_2 represent the static pressure, whereas the terms $\tfrac{1}{2}\rho v_1^2$ and $\tfrac{1}{2}\rho v_2^2$ represent the dynamic pressure. The maximum static pressure p_s is reached when the velocity term is zero and is commonly referred to as the stagnation pressure.

EXAMPLE 4-8

The racing car in Fig. 4-15 is traveling along a beach at 325 km/hr (about 200 mph). At the front, where the airstream parts, the velocity is zero. Given a density for air of 1.227 kg/m³ (0.002 38 slug/ft³), what is the maximum air pressure that can develop at the front of the car, i.e., the stagnation pressure?

SOLUTION
From Eq. (4.29)

$$p_s = p + \tfrac{1}{2}\rho v^2 = (1.01 \times 10^5 \text{ N/m}^2) + \tfrac{1}{2}(1.227 \text{ kg/m}^3)(90.27 \text{ m/s})^2$$

and

$$p_s = 1.06 \times 10^5 \text{ N/m}^2 \ (15.4 \text{ psi})$$

Notice in the equation after the plus sign that the units $\text{kg/ms}^2 = ML^{-1}T^{-2} = FL^{-2}$, which is consistent with the pressure units N/m² in the answer.

4-6   TORRICELLI'S THEOREM

Torricelli's theorem is a special case of Bernoulli's equation derived from an ideal system.

In Fig. 4-9 a large tank holds water that flows horizontally out a small pipe at the base. Bernoulli's equation for the system is

$$z_1 + \left(\frac{1.02 \times 10^{-4} p_1}{Sg} \right) + \left(\frac{v_1^2}{2g} \right) = z_2 + \left(\frac{1.02 \times 10^{-4} p_2}{Sg} \right) + \left(\frac{v_2^2}{2g} \right) + H_e + H_l$$

Since the system is considered ideal, losses are negligible and H_1 equals 0. The elevation h equals $(z_1 - z_2)$, and if z_2 is zero, h equals z_1. If the top of the tank is open to the atmosphere, the gauge pressue is 0 at the surface and p_1 equals 0. The pressure p_2 at the surface of the outlet near the tank is also 0, since the free jet is discharged into the atmosphere. Where the area of the surface of water in the tank is sufficiently large, the velocity of the surface is negligible and the velocity head term $v_1^2/2g$ also becomes 0. Bernoulli's equation for this case then becomes

$$z_1 = h = \left(\frac{v_2^2}{2g} \right)$$

and

$$v_2 = \sqrt{2gh} \qquad\qquad (4.30)$$

which is a statement of Torricelli's theorem.

EXAMPLE 4-9

Compute the velocity of a fluid flowing horizontally from the base of a water tank located 20 m (65.6 ft) below the water level.

SOLUTION
Substituting in Torricelli's formula yields

$$v_2 = \sqrt{2gh} = \sqrt{(2)(9.8 \text{ m/s}^2)(20 \text{ m})} = 19.8 \text{ m/s (65 fps)}$$

EXAMPLE 4-10

The base of the water tower in Fig. 4-16 is 20 m (65.6 ft) above the ground. If the water is standing 10 m (32.8 ft) in the tank, how far from the base of the tank will the water stream touch the ground?

SOLUTION
The horizontal velocity of the water stream at the outlet of the tank is computed by solving Torricelli's theorem.

$$v_2 = \sqrt{2gh}$$

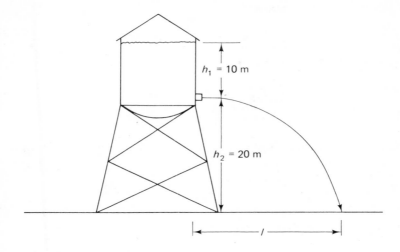

Fig. 4-16   Example 4-10

and

$$v_2 = \sqrt{(2)(9.8 \text{ m/s}^2)(10 \text{ m})} = 14 \text{ m/s } (45.9 \text{ ft/sec})$$

From the time when the water leaves the outlet at the base of the tank horizontally, it is free-falling, and the distance from the base of the tank that it will travel before striking the ground is influenced by the horizontal velocity, the force of gravity, and the height it falls before striking the ground. From Eq. (4.20), the terminal vertical velocity at the point of impact is

$$v = gt$$

and since the water began falling from a position of rest, and the acceleration due to gravity is approximately constant, the average velocity is $v/2$, and the distance through which it falls equals the average velocity × time. That is,

$$h = \left(\frac{v}{2}\right)(t)$$

and substituting gt for v gives

$$h = \tfrac{1}{2} gt^2 \qquad\qquad (4.31)$$

Solving for the elapsed time t, we obtain

$$t = \sqrt{\frac{2h}{g}} = \sqrt{\frac{(2)(20\text{ m})}{(9.8\text{ m/s}^2)}} = 2.02\text{ s}$$

Thus, the horizontal distance that the water travels from the base of the tank before striking the ground is

$$l = vt = (14\text{ m/s})(2.02\text{ s}) = 28.3\text{ m} \; (92.8\text{ ft})$$

4-7   SUMMARY AND APPLICATIONS

Flow phenomena are concerned with movement of fluids within systems' boundaries as volume is displaced from one point or place to another. Flow is considered to be ideal if the fluid is incompressible and exhibits no viscous shear. Most flow problems consider the flow to be one-dimensional, although two- and three-dimensional flows must be given consideration when their effect causes appreciable variances between the observed results, and those expected when the flow is considered to be one-dimensional. Such would be the case, for example, when the boundary of the fluid changes cross section. Fluid flow may be steady or unsteady past a fixed point, uniform or nonuniform with respect to velocity and direction along the line of flow, laminar if the flow is smooth and governed by the viscous nature of the fluid, or turbulent if agitation disrupts the predictable flow path of fluid particles, requiring empirical rather than theoretical solutions to related flow problems.

Displacement describes the working volume of fluid power machinery per cycle and is usually measured in cubic units. It differs from flow rate, which considers the displaced volume per unit time at any place in the system, for example, in a water flume or branch air line. The continuity equation relates the conduit area to the fluid velocity in such a way that the fluid mass passing any point in a continuous conduit is the same. A number of fluid power problems rely on the continuity equation as a preliminary step to size the conductor or establish the velocity of fluid within acceptable limits. The energy equation totals the potential energy from the elevation and work-flow pressure within the fluid, with the kinetic energy caused by movement of the fluid, to determine the total energy level of the fluid. Energy levels between two or more points in the system are equated to resolve one or more unknowns in the equation, or to account for variances in the potential or kinetic energy. Power added or

Fig. 4-17   Application 6 (*Courtesy of Technovate, Inc.*)

subtracted from the system as well as frictional losses are included when these affect the energy level of the system. The balance of energy between two points in the system is accounted for by Bernoulli's theorem. The aerodynamic form of the energy equation, and Torricelli's equation which governs the flow through nozzles, are special cases of the Bernoulli equation.

Following are related applications that are useful to develop several common concepts and principles that govern the flow of fluids.

1. Verify the displacement of one or more fluid power components by accurately measuring the volume pumped or consumed per cycle.
2. Calibrate a variable area liquid flowmeter, or, using a calibrated variable area flowmeter, verify the output from a direct reading flowmeter connected with it in series. Construct tables and graphs of results, comparing the characteristics of each instrument.
3. Construct a fluid system that validates the continuity equation.
4. Quantify the pressure-volume relationship of air or other gases by transferring compressed volumes from one size vessel to another, as from low-pressure tanks to balloons.
5. Verify Bernoulli's theorem, using an apparatus such as that shown in Fig. 4-12 for incompressible fluids.
6. Verify the aerodynamic form of Bernoulli's equation, using an apparatus such as that in Fig. 4-17.
7. Construct an apparatus to verify Torricelli's theorem, using the time and point of impact of a free falling horizontal jet that discharges from a raised tank or other pressurized source.

4-8   STUDY QUESTIONS AND PROBLEMS

1. What is meant by *fluid boundary*?
2. Define *one-*, *two-*, and *three*-dimensional flow.
3. What is meant by *steady conditions*? How do *steady conditions* differ from *uniform flow*?

4. Define *laminar* and *turbulent* flow.

5. What is meant by the term *displacement*?

6. A 5-cylinder radial piston pump turning at 500 rpm delivers 25 liters (6.6 gpm) of hydraulic fluid per minute. Compute the displacement of the pump and one cylinder through one revolution.

7. How does *displacement* differ from *flow rate*?

8. Fluid flows through a 2.5-cm (1-in.) diameter hydraulic line at 50 liters/min (13.2 gpm). Compute the velocity in m/s (ft/sec).

9. A 2-cylinder single-stage compressor delivers 0.25 m³/min (8.83 ft³/min) at 1 MPa (145 psi). Compute the air delivery from the compressor in m³ and ft³ of free air per minute.

10. Restate the continuity equation.

11. How large would the inside diameter of a pipe have to be to deliver 150 gpm (568 l/m) if the velocity is limited to 10 ft/s (3.05 m/s)?

12. A pipe carrying water at the flow rate of 125 liters/min (33 gpm) is reduced from 7.5 cm (3 in.) to 2.5 cm (1 in.) in diameter. Compute the velocity of the fluid in both sections of the pipe.

13. Crude oil with a Sg of 0.90 is flowing horizontally in a pipe that reduces in diameter from 15 cm (6 in.) to 7.5 cm (3 in.). In the 15-cm (6-in.) section, the pressure is 10 bars (145 psi), and the flow is 2000 liters/min (528 gpm). Neglecting friction, compute the fluid velocity in the 15-cm (6-in.) section of the pipe, and the flow rate and pressure in the 7.5-cm (3-in.) section of pipe.

14. Water is piped from a 5-cm (2-in.) diameter main operating at 1200 kPa (174 psi) upward to the seventh floor of a building in a 3-cm (1.2-in.) diameter branch line. The flow rate in the 5-cm (2-in.) main is 75 liters/min (19.8 gpm) and friction losses in the branch line are 0.05 m/m of pipe. Compute the velocity in the 5-cm (2-in.) main, and the velocity and pressure available on the seventh floor.

15. A 5-cm (2-in.) plug near the base of a cylindrical water tank with 5 m (16.4 ft) of head is removed. What will be the initial velocity of the fluid stream from the tank?

16. The base of a storage tank 50 m (164 ft) above the ground surface contains an additional 10 m (32.8 ft) of water. If two nozzles at the base of the tank, one pointed horizontally and one vertically downward, are opened simultaneously, compute the time t for each stream of water to touch the ground.

5

FLUID DYNAMICS

5-1 INTRODUCTION

Fluid dynamics considers the behavior of fluids in motion. In the present discussion, fluid flow is considered to be one-dimensional and under steady conditions, during which the control volume is translated within the boundaries of the system.

Unlike simple flow phenomena, which deal primarily with the energy balance in the system, fluid dynamics considers the work performed on or by the fluid and the power thus transmitted for useful purposes.

The transmission of useful power involves fluid friction losses. Their value and calculation are influenced by whether the fluid is noncompressible or compressible, and whether it is flowing in a laminar or turbulent mode.

5-2 FLUID WORK AND POWER

Most systems produce useful work as an objective, typically by the means that yields the highest practical efficiency. Work W is defined as the application of a force F through some distance L. In notation

$$W = FL \tag{5.1}$$

In SI units, work is measured in joules (J), where one joule equals the work performed when an object under a force of 1 N moves 1 m in that direction. In English units, work is measured in ft-lb. While work and potential energy have the same dimensions, conceptually they are different. And so, work involves motion as does kinetic energy.

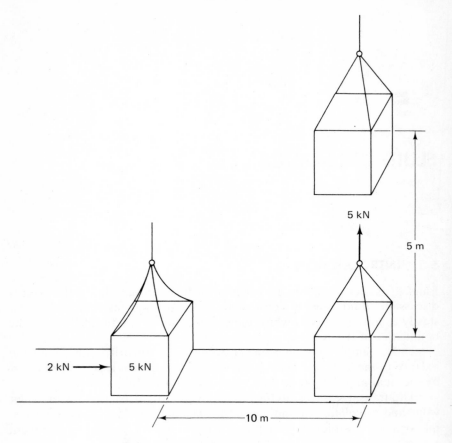

Fig. 5-1 Work performed on an objective

Figure 5-1 illustrates an example of an objective upon which work is being performed, first horizontally, and then vertically. In the horizontal direction

$$W = (2 \times 10^3 \text{ N})(10 \text{ m}) = 20 \text{ kJ } (\text{m} \cdot \text{kN} \quad or \quad 14\,800 \text{ ft-lb})^1$$

In the vertical direction

$$W = (5 \times 10^3 \text{ N})(5 \text{ m}) = 25 \text{ kJ } (\text{m} \cdot \text{kN} \quad or \quad 18\,500 \text{ ft-lb})$$

Power P is the rate at which work is accomplished, for example, by a wind

---

[1]The conversion from joules to ft-lb is: 1 joule $= 1$ m$\cdot$N $= (1$ m$\cdot$N)(3.28 ft/m)(1/4.448 lb/N) $= 0.74$ ft-lb, and 1 ft-lb $= 1.36$ J.

current due to the movement of an air mass, the fall of water from a dammed source, or the flow of fluid through pipes under pressure. That is,

$$P = \frac{FL}{t} = \frac{W}{t} \tag{5.2}$$

where t is the time in seconds and P is in joules/second (watts) or ft-lbf/sec.

The horsepower is a convenient unit attributed to James Watt (1736–1819), who, needing a standard to measure the power of his steam engines, determined experimentally the capability of draft horses to lift weights suspended across a pully. The value of the horsepower was fixed arbitrarily at 550 ft-lb/sec (746 J/s). This has been the standard for rotating machinery since that time. In SI units, 1 watt = 1 joule/second and the horsepower equals

$$HP = \frac{P}{746} \tag{5.3}$$

In English units

$$HP = \frac{FL}{t\,550} \tag{5.4}$$

Horsepower is a number without dimension.

Fluid power components such as cylinders and motors receive and transmit power in the form of fluid volume flow Q under pressure p through pipes. Since $(p)(A) = F$ and the distance L is equated to the stroke s, the power from a hydraulic cylinder CP can be computed from

$$CP = \frac{pAs}{t} \tag{5.5}$$

And since

$$Q = \frac{(A)(s)}{t}$$

the fluid power FP through the system equals

$$FP = pQ \tag{5.6}$$

In SI units, cylinder horsepower CHP and fluid horsepower FHP are computed by using

$$CHP = \frac{(p \times 10^5 \text{ N/m}^2)(A \times 10^{-4} \text{ m}^2)(s \times 10^{-2} \text{ m})}{(t \text{ sec})(746 \text{ m} \cdot \text{N/s})} = \frac{pAs}{7460t} \tag{5.7}$$

and

$$FHP = \frac{(p \times 10^5 \text{ N/m}^2)(Q \times 10^{-3} \text{ m}^3/\text{min})(1/60 \text{ min/sec})}{(746 \text{ m} \cdot \text{N/s})}$$

$$= \frac{pQ}{448} \tag{5.8}$$

where pressure is given in bars and bore and stroke are given in cm, and the flow rate is in l/m.

In English units, cylinder horsepower and fluid horsepower are computed by using

$$CHP = \frac{(p \text{ lbf/in}^2)(A \text{ in}^2)(s \text{ in.})(1/12 \text{ ft/in.})}{(t \text{ sec})(550 \text{ ft-lbf/sec})} = \frac{pAs}{6600t} \tag{5.9}$$

and

$$FHP = \frac{(p \text{ lbf/in}^2)(Q \text{ gal/min})(1/60 \text{ min/sec})(231 \text{ in}^3/\text{gal})(1/2 \text{ ft/in.})}{(550 \text{ ft-lbf/sec})} \tag{5.10}$$

$$FHP = \frac{pQ}{1714}$$

where the pressure is in lbf/in² and bore and stroke are in inches; the flow rate is in gpm.

Rotating machinery imparts or receives a turning moment or torque at the shaft. From the general power formula (5.2) and Fig. 5-2, it is seen that the force component F acts at a tangent to the shaft, i.e., at right angles to the radius through the distance $(D/2)$, whereas the velocity component equals the circumferential rotation of the shaft in revolutions per second (rps). That is,

$$\frac{L}{t} = \frac{(C)(n)}{(t)} = \frac{(\pi)(D)(n)}{(t)}$$

where C equals the circumference of the shaft and n equals the total number of rotations. Torque power (TP) equals

$$TP = \frac{(F)(\pi)(D)(n)}{(t)}$$

But $n/t = N/60$, where N is revolutions per minute (rpm), and $(D)(F) = 2T$, where T is the torque in N·m (lbf·ft). So that

$$TP = \frac{2\pi NT}{60} \tag{5.11}$$

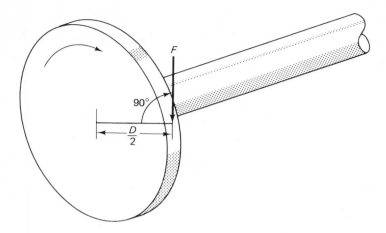

Fig. 5-2 Torque as a twisting moment

In SI units, torque horsepower is computed from

$$THP = \frac{\dfrac{2\pi NT}{60}}{746} = \frac{2\pi NT}{44\ 760} \qquad (5.12)$$

In English units

$$THP = \frac{\left(\dfrac{2\pi NT}{60}\right)}{550} = \frac{2\pi NT}{33\ 000} \qquad (5.13)$$

EXAMPLE 5-1

Water with a head of 50 m (164 ft) from a reservoir feeds a turbine at the rate of 250 m³/s (8829 ft³/sec). Neglecting friction losses, compute the available power in SI and English units, and the horsepower.

SOLUTION
In SI units, the power available from Eq. (5.2) is

$$P = \frac{FL}{t} = \frac{(9802\ \text{N/m}^3)(250\ \text{m}^3)(50\ \text{m})}{(1\ \text{s})} = 122.5\ \text{MW (megawatts)}$$

In English units, the power available is

$$P = \frac{(62.4\ \text{lb/ft}^3)(250\ \text{m}^3 \times 35.31\ \text{ft}^3/\text{m}^3)(50\ \text{m} \times 3.2808\ \text{ft/m})}{(1\ \text{s})}$$

and

$$P = 9.036 \times 10^7 \text{ ft-lb/sec}$$

The horsepower potential is

$$HP = \frac{P}{746} = \frac{(122.5 \text{ MW})}{(746 \text{ W/HP})}$$

$$= 164 \text{ kHP (kilo-horsepower)}$$

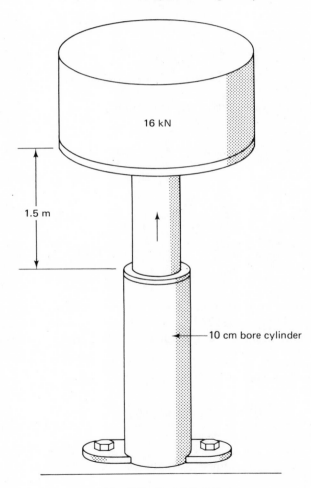

Fig. 5-3   Example 5-2

EXAMPLE 5-2

A fluid power cylinder with a 10-cm (4-in.) bore raises a mass of 1635 kg (3597 lb) through 1.5 m (4.9 ft) in 4 sec (Fig. 5-3). Compute the system pressure and cylinder horsepower.

SOLUTION
From Fig. 5-4

$$p = \frac{F}{A} = \frac{(4)(1635 \text{ kg})(9.8 \text{ m/s}^2)}{(3.14)(10 \times 10^{-2} \text{ m})^2} = 20.4 \times 10^5 \text{ Pa} = 20.4 \text{ bars (296 psi)}$$

And from Eq. (5.7)

$$CHP = \frac{pAs}{7460t} = \frac{(20.38 \text{ bars})(78.5 \text{ cm}^2)(150 \text{ cm})}{(7460)(4 \text{ s})} = 8.0 \text{ hp}$$

EXAMPLE 5-3

A hydraulic system is equipped with a pump that delivers 20 l/min (5.28 gpm) at a pressure of 135 bars (1957.5 psi). Compute the fluid horsepower potential of the system.

SOLUTION
From Eq. (5.8)

$$FPH = \frac{pQ}{448} = \frac{(135 \text{ bars})(20 \text{ l/m})}{(448)} = 6.01 \text{ hp}$$

EXAMPLE 5-4

Neglecting losses, compute the torque from a 12-hp fluid power motor turning at 1500 rpm.

SOLUTION
Solving Eq. (5.12) for T, we obtain

$$T = \frac{(THP)(44\,760)}{2\pi N} = \frac{(12 \text{ hp})(44\,760 \text{ N} \cdot \text{m})}{(2)(3.14)(1500 \text{ rpm})}$$

$$= 57 \text{ N} \cdot \text{m } (42 \text{ lbf-ft})$$

5-3   NONCOMPRESSIBLE FLOW IN PIPES

In circular pipes where the flow is governed by the viscous nature of noncompressible fluids, the laminae of unit thickness assume the configuration of thin shell concentric tubes sliding one over another successively,

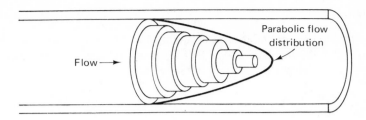

Fig. 5-4   Viscous flow

with viscous friction between them dissipating the potential energy and returning it to the fluid and system as heat. It must also be assumed that steady conditions prevail, i.e., that the velocity and volume of fluid flowing past a fixed point in the conduit are constant. Figure 5-4 illustrates the parabolic distribution of these thin shell tubes, which telescope as they flow in a circular conduit, reflecting the increase in fluid velocity near the center of the fluid stream, even though each has a velocity that is constant throughout its length.

    The basic equation that governs viscous noncompressible flow in pipes is

$$Q = \frac{\pi r^4 \rho g h}{8 L \mu} \tag{5.14}$$

where ρ equals the mass density of the fluid, r equals the internal radius of the pipe, and μ equals the absolute viscosity of the fluid. Since the flow rate Q can be equated to the cross-section area of the pipe times the fluid velocity, i.e., $Q = (A)(v) = (\pi r^2)(v)$,

$$\pi r^2 v = \frac{\pi r^4 \rho g h}{8 L \mu}$$

Solving for the head loss associated with producing viscous flow through a pipe of length L, we obtain

$$h_f = \frac{8 \mu L v}{\rho g r^2} \tag{5.15}$$

Typically the diameter D of the pipe is given rather than the radius r and

$$h_f = \frac{32 \mu L v}{\pi D^2 g} \tag{5.16}$$

which is known as the Darcy-Weisbach formula for laminar flow.

5-4   REYNOLDS NUMBER

Osborne Reynolds[2,3] (1842-1912) discovered that viscous flow was related to the dimensionless ratio N_R of the product of the fluid velocity v, pipe diameter D, and fluid density ρ, to the absolute viscosity μ in the liquid such that

$$N_R = \frac{vD\rho}{\mu} \tag{5.17}$$

Reynolds number N_R can be substituted in Eq. (5.16) by modifying it with the expression v/v. That is,

$$h_f = \frac{(32\,\mu Lv)(v)}{(\rho D^2 g)(v)} = \left[\frac{32}{\dfrac{vD\rho}{\mu}}\right]\left(\frac{Lv^2}{Dg}\right) = \frac{32Lv^2}{N_R Dg} \tag{5.18}$$

where N_R is dimensionless and L, v^2, D, and g are in consistent units.

Reynolds apparatus (Fig. 5-5) passed water through horizontal tubes of different diameters from 1/4 to 2 inches at increasing velocities to ascertain the velocity at which a dye bled into the stream would indicate a transition from laminar to turbulent flow. This work, which began about 1880, and related work by other investigators which replicated similar experiments through 1910, indicated upper and lower critical limits for the change from laminar to turbulent flow to lie between N_R of 2000 and 4000. Below 2000 the flow is laminar, whatever its previous state, and this is considered to be the lower critical limit. Above 4000 the flow becomes or remains turbulent. The critical velocity, then, is considered to be $2000 \leqslant N_R \leqslant 4000$.

The value of the friction factor in the Darcy-Weisbach formula is largely determined by whether the flow is laminar or turbulent. For laminar flow conditions, the viscous friction factor has been determined experimentally to be

$$f = \frac{64}{N_R}$$

---

[2]For a detailed accounting of Osborne Reynolds' experiment, see G. A. Tokaty, *A History and Philosophy of Fluidmechanics*. Henley-on-Thames, Oxfordshire: G. T. Foulis and Co. Ltd., 1971.

[3]Or see Reynolds' original paper, "An Experimental Investigation of the Circumstances Which Determine Whether the Motion of Water Shall be Direct or Sinuous, and the Law of Resistance in Parallel Channels," *Philosophical Transactions of the Royal Society*, **174**, Part III, 935 (1883), or *Scientific Papers*, London: Cambridge University Press, 1900–1903, Vol. II, pp. 51–105.

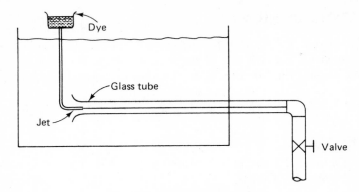

Fig. 5-5   Reynolds apparatus

Substituting this value in the modified Darcy-Weisbach formula (5.18) for head loss due to friction results in one form of the Hagen-Poiseuille formula for losses due to friction when laminar flow prevails. That is,

$$h_f = \left(\frac{64}{N_R}\right)\left(\frac{Lv^2}{D2g}\right) \qquad (5.19)$$

This indicates that for laminar flow, head losses from friction are inversely proportional to the Reynolds number and diameter of the pipe, and directly proportional to the length of the pipe and square of the velocity. When turbulent conditions prevail, the Darcy-Weisbach formula becomes

$$h_f = f\left(\frac{Lv^2}{D2g}\right) \qquad (5.20)$$

where the friction factor f is determined experimentally.

EXAMPLE 5-5

Crude oil with a Sg of 0.86 and μ of 6×10^{-3} N·s/m² (12.5×10^{-5} lbf·sec/ft²) is transferred through a 5-cm diameter pipe at the flow rate of 250 liters/min (66 gpm). Is the flow laminar or turbulent?

SOLUTION

The flow is laminar if $N_R \leqslant 2000$, and turbulent if $N_R \geqslant 4000$. The flow velocity component is computed from

$$v = \frac{Q}{A} = \frac{(4)(250 \times 10^{-3} \text{ m}^3/\text{min})(1/60 \text{ min/sec})}{(3.14)(5 \times 10^{-2} \text{ m})^2} = 2.12 \text{ m/s}$$

Since the Sg rather than ρ is given,

$$\gamma = \rho g$$

$$\rho = \frac{\gamma}{g}$$

$$\rho = \frac{(\gamma_{std})(Sg)}{g} = \frac{(9802 \text{ N/m}^3)(0.86)}{(9.8 \text{ m/s}^2)} = 860 \text{ N} \cdot \text{s}^2/\text{m}^4$$

Finally,

$$N_R = \frac{vD\rho}{\mu} = \frac{(2.12 \text{ m/s})(5 \times 10^{-2} \text{ m})(860 \text{ N} \cdot \text{s}^2/\text{m}^4)}{(6 \times 10^{-3} \text{ N} \cdot \text{s}/\text{m}^2)} = 15\ 193$$

and the flow is turbulent.

EXAMPLE 5-6

Given an oil with a Sg of 0.90 and an absolute viscosity of 27 cP $(5.6 \times 10^{-4} \text{ lbf} \cdot \text{sec}/\text{ft}^2)$ flowing in a 2.5-cm (1-in.) pipe, compute the critical velocity range, i.e., $2000 \leqslant N_R \leqslant 4000$.

SOLUTION
From Reynolds formula

$$N_R = \frac{vD\rho}{\mu}$$

and

$$v = \frac{N_R \mu}{D\rho}$$

The absolute viscosity in cP is converted to $\text{N} \cdot \text{s}/\text{m}^2$ by using

$$cP \times 10^{-3} = \text{N} \cdot \text{s}/\text{m}^2$$

and

$$\mu = 27 \text{ cP} \times 10^{-3} = 27 \times 10^{-3} \text{ N} \cdot \text{s}/\text{m}^2$$

Since the Sg rather than ρ is given,

$$\rho = \frac{\gamma_{std} Sg}{g} = \frac{(9802 \text{ N/m}^3)(0.90)}{(9.8 \text{ m/s}^2)} = 900 \text{ N} \cdot \text{s}^2/\text{m}^4$$

From which the upper velocity limit where $N_R = 4000$ is equated

$$v = \frac{(4000)(27 \times 10^{-3} \text{ N·s/m}^2)}{(2.5 \times 10^{-2} \text{ m})(900 \text{ N·s}^2/\text{m}^4)} = 4.8 \text{ m/s} \ (15.75 \text{ ft/sec})$$

and the lower velocity limit where $N_R = 2000$ is equated

$$v = \frac{(2000)(27 \times 10^{-3} \text{ N·s/m}^2)}{(2.5 \times 10^{-2}\text{m})(900 \text{ N·s}^2/\text{m}^4)} = 2.4 \text{ m/s} \ (7.88 \text{ ft/sec})$$

EXAMPLE 5-7

Liquid with a Sg of 0.85 is transferred between two tanks 300 m (984 ft) apart through a 6.5-cm (2.6 in) pipe at 1000 liters/min (264 gpm). If the friction factor is given as 0.05, what is the pressure drop between the tanks?

SOLUTION

With reference to Fig. 5-6, the velocity of the fluid through the pipe is

$$v = \frac{Q}{A} = \frac{(4)(1000 \times 10^{-3} \text{ m}^3/\text{min})(1/60 \text{ min/sec})}{(3.14)(6.5 \times 10^{-2} \text{ m})^2} = 5.03 \text{ m/s}$$

Substituting in Eq. (5-20) to solve for head losses when the flow is turbulent, we obtain

$$h_f = f\left(\frac{Lv^2}{D2g}\right) = \frac{(0.05)(300 \text{ m})(5.03 \text{ m/s})^2}{(6.5 \text{ m})(2)(9.8 \text{ m/s}^2)} = 2.98 \text{ m} \ (3.6 \text{ psi})$$

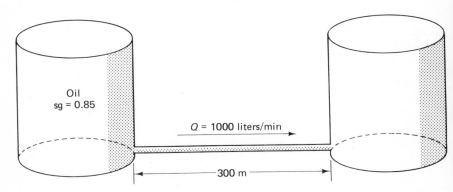

Oil
sg = 0.85

Q = 1000 liters/min

300 m

Fig. 5-6   Example 5-7

5-5   COMPRESSIBLE FLOW

Unlike noncompressible flow in pipes, where the ρ of the liquid remains relatively constant, compressible flow is accompanied by reduction in ρ as the pressure drops. Reviewing the continuity equation of mass flow in Eq. (4.13)

$$\rho_1 A_1 v_1 = \rho_2 A_2 v_2$$

where

$$\rho = \frac{\gamma}{g}$$

and

$$\gamma = \frac{p}{h}$$

Then by substitution

$$\rho = \frac{p}{gh}$$

Substituting in the continuity equation, we have

$$\frac{p_1 A_1 v_1}{gh_1} = \frac{p_2 A_2 v_2}{gh_2} \qquad \textbf{(5.21)}$$

For flow through horizontal pipes with no change in elevation h

$$p_1 A_1 v_1 = p_2 A_2 v_2 \qquad \textbf{(5.22)}$$

This indicates that ρ varies directly with the pressure p. If the absolute pressure drops to one-half its original value so does the ρ of the fluid, given, of course, that the temperature remains constant. It also shows that for a conductor of constant cross section, the continuity equation will require the velocity to double. Typically, what happens in practice is that large changes in pressure are accompanied by a change in both cross-section and velocity, for example, as the fluid passes through a venturi, tapered tube, or ventilating duct reducer.

At low pressures and flow rates, for example, in ventilating systems with less than 10 percent pressure drop, compressible flow and noncompressible flow are similar. The major difference is the density of the compressible gas as compared with that of an incompressible liquid, such as water. Underlying this assumption is that the flow is steady and that the pressure drop throughout the length of the conductor is not sufficient to

Fig. 5-7   Anemometer

change the density of the compressible fluid. In practical applications, friction losses thus can be computed by using a reference fluid such as water, multiplied by a density correction factor. Figure 5-7 illustrates an anemometer for measuring the velocity of gases through ventilating systems.

It is also assumed that the flow is one-dimensional, that the fluid follows the laws for an ideal fluid, that changes in temperature are negligible, and that pressure gradients do not exist throughout the length of the conductor. In practice minor violations of these assumptions do not appreciably affect the accuracy and usefulness of computed results.

The relationship between the density and pressure of a reference fluid such as water, and the fluid flowing such as air, is

$$\frac{(\rho_{\text{ref. fld.}})}{(\rho_{\text{gas}})} = \frac{(h_{\text{gas}})}{(h_{\text{ref. fld.}})}$$

$$(\rho_{\text{ref. fld.}})(h_{\text{ref. fld.}}) = (\rho_{\text{gas}})(h_{\text{gas}})$$

and

$$h_{\text{ref. fld.}} = (h_{\text{gas}})\frac{(\rho_{\text{gas}})}{(\rho_{\text{ref. fld.}})}$$

But $h_{\text{gas}} = h_f$ in Eq. (5.20). Substituting, we have

$$h_{\text{ref. fld.}} = f\left(\frac{Lv^2}{D2g}\right)\left(\frac{\rho_{\text{gas}}}{\rho_{\text{ref. fld.}}}\right) \qquad (5.23)$$

EXAMPLE 5-8

A round smooth ventilating duct 25 cm (10 in.) in diameter delivers air at 8 m³/min (283 ft³/min) at 20°C (68°F) and atmospheric pressure. If the friction is taken as $f = 0.015$, determine the pressure drop per 100 m (328 ft) of duct.

SOLUTION

The velocity of the fluid is determined from

$$v = \frac{Q}{A} = \frac{(4)(8 \text{ m}^3/\text{min})(1/60 \text{ sec}/\text{min})}{(3.14)(25 \times 10^{-2} \text{ m})^2} = 2.7 \text{ m/s}$$

At 20°C, the density of air can be taken as $1.2 \text{ kg/m}^3 (2.3 \times 10^{-3}$ slugs/ft³), and the density of water as 998 kg/m³. Substituting in Eq. (5.23), we obtain

$$h_{\text{ref. fld.}} = (0.015)\frac{(100 \text{ m})(2.7 \text{ m/s})^2(1.2 \text{ kg/m}^3)}{(25 \times 10^{-2})(2)(9.8 \text{ m/s}^2)(998 \text{ kg/m}^3)}$$

$$= 2.683 \times 10^{-3}\text{m } (2683\mu)$$

or

$$p_{\text{ref.fld.}} = (9802 \text{ N/m}^3)(2.68 \times 10^{-3} \text{ m})$$

$$= 26.3\text{Pa}$$

As a fraction, this represents only about $(26.3 \text{ N/m}^2)/(1.01 \times 10^5$ N/m²) $= 2.6 \times 10^{-4}$ of the pressure at the entrance of the pipe, indicating this method of computing head loss would be suitable for a much longer section of duct.

While the friction factor was given rather than computed from (N_R) in Example 5-8, it should be noted that the Reynolds number does not change appreciably along the pipe length, since the product $v\rho$ is constant for a perfect gas and μ varies only slightly with changes in pressure.

5-6 SUMMARY AND APPLICATIONS

The dynamics of fluid flow usually result from activity designed to accomplish work and transmit power. Work is accomplished by a number of means, depending upon the system design. Figure 5-8 illustrates a flow diagram of a hydraulic system that transmits power from a pump to a cylinder; the figure indicates that the system output objective is transposed into system input and energy source requirements. The output horsepower consists of force, length, and time factors, which are transposed from cylinder horsepower factors of pressure, area, length, and time to the fluid horsepower factors of pressure and volume flow rate.

The Reynolds number, sometimes called the flow index, consists of the dimensionless ratio of the product of the fluid velocity, pipe diameter, and fluid density, to the absolute viscosity of the fluid. For values $N_R \leqslant 2000$ the flow has been found to be laminar. For values $N_R \geqslant 4000$ the flow is turbulent. Flow becomes unstable when $2000 \leqslant N_R \leqslant 4000$, and this is called the critical range. For values of $N_R \leqslant 2000$, the friction factor associated with fluid flow has been determined experimentally to be $f = 64/N_R$.

The behavior of noncompressible flow in circular pipes is governed by whether the flow is laminar or turbulent. Where the flow is laminar and $N_R \leqslant 2000$, flow losses are proportional to the absolute viscosity, pipe length, and velocity, and inversely proportional to the density of the fluid

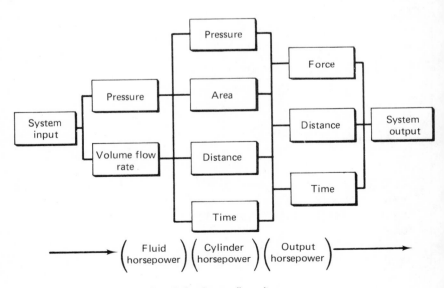

Fig. 5-8 Power flow diagram

and square of the diameter. The friction factor can be determined from N_R. Turbulent flow, i.e., where $N_R \geqslant 4000$, is not governed by the viscous nature of the fluid and the friction factor is usually determined experimentally.

Compressible flow is accompanied by changes in ρ with respect to pressure. For low velocities where the pressure drop is less than 10 percent throughout the length of round duct in question, it can be assumed that the density does not change and that friction losses can be computed by using a reference fluid such as water, multiplied by a density correction factor, without introducing appreciable error.

Following are related applications that are useful to develop several common concepts and principles that govern the dynamics of fluids.

1. Discharge a positive displacement hydraulic pump through a flow control valve and flow meter. Monitor both pressure and flow rate, and use the fluid horsepower formula to verify the output horsepower with respect to changes in pump pressure.

2. Compare water or hydraulic pump input and output power data with respect to pressure. Graph both curves.

3. Using a positive displacement pump connected to a cylinder under load, verify the relationship between fluid horsepower and cylinder horsepower.

4. Using a friction dynamometer, plot the torque curve for a motor or turbine.

5. Demonstrate laminar and turbulent flow phenomena (see footnotes 2 and 3) or, using the references noted and Fig. 5-6, construct a Reynolds apparatus to demonstrate laminar and turbulent flow.

6. Compute the Reynolds number for different flow velocities, verifying that for $N_R \leqslant 2000$ the flow is laminar and for $N_R \geqslant 4000$ the flow is turbulent.

7. Using an anemometer such as that in Fig. 5-7 to record air velocity, compute the volume flow rate and weight flow rate of air through a round or square ventilating duct.

5-7 STUDY QUESTIONS AND PROBLEMS

1. A single-acting automotive lift raises a car weighing 10 kN (2248 lb) 2 m (6.56 ft). If it is assumed that the lift itself weighs 2.225 kN (500 lb) and that friction is negligible, what is the work necessary to raise the car in joules and ft-lb?

2. From Problem 1, construct a table indicating the horsepower necessary to raise the car in 5, 10, 15, 20, 25, and 30 seconds.

3. From Problem 1, compute the bore of the cylinder if the system pressure is 7000 kPa (1015 lbf/m^2) and the time is 8 seconds.

4. A hydraulic pump delivers 45 l/min (11.9 gpm) at a pressure of 103 bars (1494 lbf/m^2). Compute the potential fluid horsepower of the system.

5. A 5-hp motor drives a hydraulic pump that delivers fluid at a pressure of 14 MPa (2030 lbf/m^2). Assuming no losses, compute the flow rate in l/min and gpm.

6. Compute the motor torque in Problem 5 in $N \cdot m$ and $lbf\text{-}ft$.

7. A sump pump lifts water 10 m (32.8 ft) through a 2.5-cm (1-in.) diameter pipe at 40 l/min (10.6 gpm). Assuming no losses, compute the horsepower required.

8. Water feeds a turbine at the base of a dam at 150 m^3/s (5297 ft^3/sec). If the water surface is 70 m (230 ft) above the turbine and friction losses are ignored, what is the potential power in SI and English units, and the horsepower?

9. A 7.5-cm (3-in.) diameter jet nozzle directs a free water jet against the buckets of a turbine wheel. If the jet is below the dam and fed by water with a head of 50 m, (164 ft), and friction losses are 25 percent, what is the horsepower available from the turbine?

10. Water at room temperature is flowing through a 7.5-cm (3-in.) diameter pipe at 2500 l/min (660 gpm). Compute the Reynolds number.

11. Gasoline with a Sg of 0.68 and a μ of 2.5×10^{-4} $N \cdot s/m^2$ (5.2×10^{-6} $lbf \cdot sec/ft^2$) is pumped through a 10-cm (4-in.) line at 500 l/min (132 gpm). Is the flow laminar or turbulent?

12. Given the fluid and pipe line size in Problem 10, compute the critical range of velocities, i.e., $2000 \leqslant N_R \leqslant 4000$.

13. Liquid with a Sg of 0.95 and μ of 6×10^{-3} $N \cdot s/m^2$ (12.5×10^{-5} $lbf \cdot sec/ft^2$) is to be pumped from one holding tank to another through a 5-cm (2-in.) pipe. Compute the maximum flow rate such that the friction factor is determined by using $N_R = 2000$.

14. At what maximum velocity can the flow of crude oil at 40°C (104°F) through a 2.5-cm (1-in.) diameter pipe be expected to be laminar?

15. Crude oil with a Sg of 0.68 is pumped horizontally from plant A to plant B, located 16 km (52 493 ft) away, through a 45-cm (17.7-in.) diameter pipe line at 4000 l/min (1057 gpm). If the terminal pressure is to be 2000 kPa (290 lbf/m^2) and total losses including friction losses through the pipe are estimated to be 0.5 m (1.64 ft) per 100 m (328 ft) of pipeline, compute the horsepower of the pump located at plant A.

6

RESISTANCE TO FLOW

6-1 INTRODUCTION

Resistance to fluid flow causes major and minor pressure or head losses, which subsequently result in flow and horsepower losses. Major losses are those associated with the flow of fluid through the length of pipe or channel itself and usually account for most of the total loss. Minor losses occur as the fluid negotiates a path through sudden contractions or enlargements, bends, fittings, and valves and account for a lesser part of the total loss, unless, of course, the circuit length is short and incorporates several fittings. Where the friction factor f in the Darcy-Weisbach formula is influenced by the viscosity of the fluid in laminar flow, it is influenced primarily by the relative roughness of the boundary surface in turbulent flow. For straight pipes and turbulent flow, the friction factor and major pressure losses are computed from the relative roughness, pipe diameter and Reynolds number. The friction resulting in minor losses such as when a fluid flows through a valve, fitting, or orifice is computed from a K-value, which has been determined experimentally for each fitting and flow condition, and then converted to an equivalent length of straight pipe. Total head or pressure losses for a circuit are computed by adding the major and minor losses and then substituting them in one of the formulas to determine the horsepower associated with pumping the fluid.

6-2 TURBULENT FLOW

The friction from flow in pipes brings to mind the notion that the fluid rubs the boundary surface as if a close-fitting solid body were being pushed through a tube generating heat losses at the boundary. This is not the case. What occurs is a rubbing action between the fluid particles themselves. When the flow is laminar, fluid layers move one past another, generating friction that can be related to the viscosity of the fluid. At the boundary the fluid velocity is zero; at the center of the fluid stream the velocity is maximum; and through the stream the friction gradient is linear.[1] The transition to turbulent flow introduces another important cause for friction, and that is the cross flow and intermingling of particles as the total mass moves through the length of the pipe. The effect of this intermingling of flowing particles is to void the relationship between the viscosity of the fluid and the friction generated, which then is related to the relative roughness of the boundary surface and the Reynolds number. Again, whereas it is recognized that the friction is not between the flowing fluid and the wall, the rate of shear and the heat thus generated are, in fact, greatest near the wall, and here is where most of the energy transfer occurs. At the pipe wall itself, since there is no fluid movement, no work is performed or mechanical energy dissipated. There is a heat loss, however.

6-3 f-FACTOR

When the flow is completely turbulent, the f-factor is read from Fig. 6-1 by locating the place of intersection of the Reynolds number and the relative roughness, and then reading the friction value from the left or right margins. The relative roughness of the pipe wall is computed as the dimensionless ratio of the absolute roughness to the diameter. That is,

$$\text{Relative roughness} = \frac{\epsilon}{D} \qquad (6.1)$$

where the absolute roughness equals the average projection of the surface imperfections on the inside wall of the pipe. The absolute roughness of pipes made from different materials is not the same. For example, drawn brass, copper, and glass tubing is smooth compared to pipe made of concrete and riveted steel. Table 6-1 lists the absolute roughness of several types of commercially available pipe and tubing. It must be remembered that the values shown are for new surface conditions and do not represent the surfaces of the pipes after they have been in use for some time. The procedure for solving several of the problems typically encountered is as follows:

1. Determine the fluid velocity through the pipe.
2. Compute the Reynolds number to confirm that the flow is completely turbulent.

---

[1] This is strictly true only for Newtonian fluids.

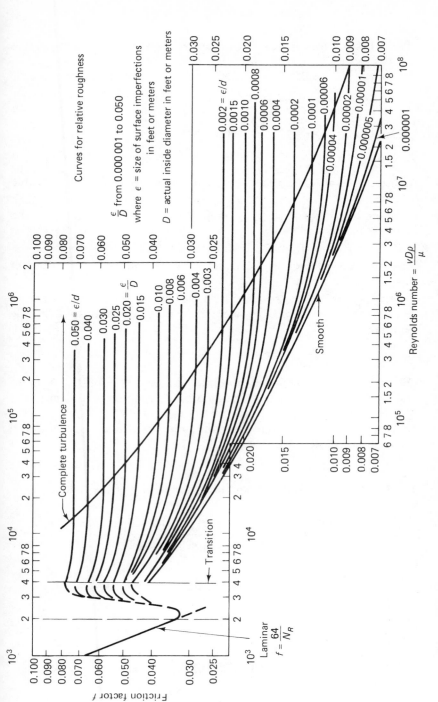

Fig. 6-1 Friction factor vs. Reynolds number for turbulent flow (*Reprinted by permission of McGraw-Hill Book Company*)

TABLE 6-1   Absolute roughness of new commercially available pipe and tubing

	ε in Feet	ε in Meters
drawn tubing	0.000 005	0.000 001 5
commercial steel pipe	0.000 15	0.000 046
asphalted cast iron	0. 0004	0.000 12
galvanized iron	0.0006	0.0002
cast iron	0. 00085	0. 000 26
concrete	0.001 to 0.01	0.000 30 to 0.0030
riveted steel	0.003 to 0.03	0.000 91 to 0.0091

3. Ascertain the absolute roughness of the pipe from the data in Table 6-1.

4. Compute the relative roughness from the absolute roughness and the diameter of the pipe.

5. Locate the f-factor from Fig. 6-1, using the Reynolds number.

6. Substitute the f-factor in the Darcy-Weisbach formula for turbulent flow and solve for h_f.

7. Determine the power associated with the friction or total loss. In cases where total losses are to be computed, it is advisable to write the Bernoulli equation.

EXAMPLE 6-1

Oil with a Sg of 0.85 and absolute viscosity of 7×10^{-3} N·s/m^2 $(1.46 \times 10^{-4}$ lbf·sec/ft$^2)$ is flowing horizontally through a 5-cm (2-in.) diameter commercial steel pipe at the rate of 400 liters/min (106 gpm). Compute the pressure drop and horsepower loss associated with the friction in 200 m (656 ft).

SOLUTION

1. From the continuity equation the velocity of the fluid is

$$v = \frac{Q}{A} = \frac{(4)(400 \times 10^{-3}\text{m}^3/\text{min})(1/60 \text{ min/sec})}{(3.14)(5 \times 10^{-2}\text{m})^2} = 3.4 \text{ m/s}$$

2. Computing the Reynolds number, we have

$$N_R = \frac{vD\rho}{\mu}$$

$$\rho = \frac{\gamma_{\text{std}} \times Sg}{g} = \frac{(9802 \text{ N/m}^3)(0.85)}{(9.8 \text{ m/s}^2)} = 850 \text{ N·s}^2/\text{m}^4$$

and

$$N_R = \frac{(3.4 \text{ m/s})(5 \times 10^{-2} \text{ m})(850 \text{ N} \cdot \text{s}^2/\text{m}^4)}{(7 \times 10^{-3} \text{ N} \cdot \text{s/m}^2)} = 20\,643$$

and the flow is turbulent.

3. From Table 6-1 the absolute roughness of commercial steel pipe is 0.000 046 m.
4. The relative roughness is computed as

$$\text{Relative roughness} = \frac{(0.000\ 046 \text{ m})}{(0.05 \text{ m})} = 0.0009$$

5. From Fig. 6-1, reading across the bottom to locate $N_R =$ 20643 and up the right margin to locate a relative roughness of 0.00098, we follow the curve to the left and upward to the place where it crosses the Reynolds number. The friction factor is then read from the left margin as 0.027.
6. Substituting in the Darcy-Weisbach equation (5.20) for turbulent flow, we obtain

$$h_f = f\left(\frac{Lv^2}{D2g}\right) = \frac{(0.027)(200 \text{ m})(3.4 \text{ m/s})^2}{(5 \times 10^{-2} \text{m})(2)(9.8 \text{ m/s}^2)} = 63.7 \text{ m}$$

or

$$h_f = \gamma h \, \text{Sg} = (9802 \text{ N/m}^3)(63.7 \text{ m})(0.85)$$

$$= 5.3 \times 10^5 \text{ Pa} = 5.3 \text{ bars (77 psi)}$$

7. Finally determining the power associated with the friction loss from Eq. (5.8), we have

$$FHP = \frac{PQ}{448} = \frac{(5.3 \text{ bars})(200 \text{ l/min})}{448} = 2.4 \text{ hp}$$

6-4   *K*-VALUES

Minor losses that occur as the fluid undergoes sudden expansions or contractions, or as the fluid flows through pipe fittings, valves, and bends are usually computed as a percentage of the kinetic energy term in the

Bernoulli equation. When turbulent flow prevails, the value of this percentage is determined experimentally and then assigned a K-value for that fitting or pipe configuration. Head loss is thus computed from

$$h_{ff} = K\left(\frac{v^2}{2g}\right) \tag{6.2}$$

This formula does not hold true when viscous flow prevails.

Sudden enlargement and reduction in the cross section of the pipe result in losses because the fluid must change directions abruptly, causing increased turbulence in the form of eddies near where the two are joined. These eddies that accompany sudden enlargement can be shown both theoretically and experimentally to generate a K-value[2] such that

$$K = \left(1 - \frac{D_1^2}{D_2^2}\right)^2 \tag{6.3}$$

where D_1 is the inside diameter of the smaller pipe and D_2 is the inside diameter of the larger pipe. Sudden reduction in the cross section of the pipe produces a K-value such that

$$K = 0.5\left(1 - \frac{D_1^2}{D_2^2}\right) \tag{6.4}$$

where D_1 and D_2 are the inside diameters of the smaller and larger pipes, respectively. Figures 6-2 and 6-3 illustrate these configurations and list K-values associated with various diameter ratios.

K-values for pipe fittings, valves, and bends are determined empirically for each configuration by manufacturers and independent researchers, and have been found to be relatively independent of size. The value of K can be computed from

$$K = f_t\left(\frac{L}{D}\right) \tag{6.5}$$

where f_t equals the friction factor in the completely turbulent range and L/D is the ratio of the length of the fitting to its inside diameter. Figure 6-4 illustrates several of the more common configurations, equivalent

[2]*Flow of Fluids Through Valves, Fittings and Pipe*, Technical Paper No. 410, Chicago: Crane Company, 1976, p. 2–11.

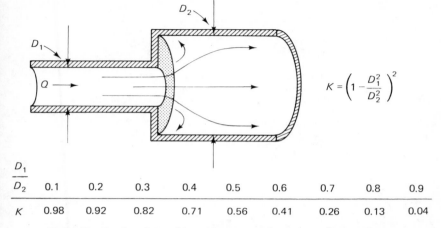

$$K = \left(1 - \frac{D_1^2}{D_2^2}\right)^2$$

$\dfrac{D_1}{D_2}$	0.1	0.2	0.3	0.4	0.5	0.6	0.7	0.8	0.9
K	0.98	0.92	0.82	0.71	0.56	0.41	0.26	0.13	0.04

Fig. 6-2 *K*-values for sudden enlargements for various diameter ratios

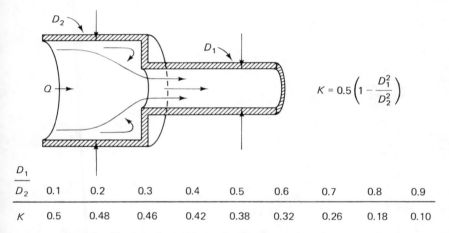

$$K = 0.5\left(1 - \frac{D_1^2}{D_2^2}\right)$$

$\dfrac{D_1}{D_2}$	0.1	0.2	0.3	0.4	0.5	0.6	0.7	0.8	0.9
K	0.5	0.48	0.46	0.42	0.38	0.32	0.26	0.18	0.10

Fig. 6-3 *K*-values for sudden reductions for various diameter ratios

lengths, and typical *K*-values associated with their length-to-diameter ratios.[3]

Losses due to expansion and contraction also occur as fluid enters and exits from pipes—for example, where they connect to the base of liquid tanks and gas receivers. Fig. 6-4 also illustrates several standard entrance and exit configurations and the *K*-values associated with each.

Following is an example that will illustrate how these losses occur in a piping system.

[3]*Ibid.*, pp. 26–29.

Valves—fittings	$\dfrac{L}{D}$	K-value
swing check valve	135	2.50
globe valve	340	10.00
gate valve		
(full open)	13	0.19
1/4 closed	35	1.15
1/2 closed	160	5.60
3/4 closed	900	24.00
cock valve	18	.26
close pattern return bend	50	2.20
standard tee	60	1.80
standard 90° elbow	30	0.90
standard 45° elbow	16	0.42

Swing check valve

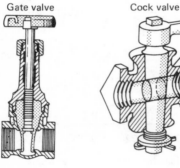

Flow in this direction only

Globe valve

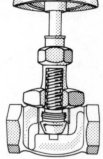

Gate valve

Cock valve

Pipe entrance

Inward projecting

r/d	K
0.00*	0.5
0.02	0.28
0.04	0.24
0.06	0.15
0.10	0.09
0.15 and up	0.04

K = 0.78   *Sharp-edged

Flush

For K, see table

Close pattern return bend

Standard tee

K = 0.78

Pipe exit

Projecting   Sharp-edged   Rounded

K = 1.0   K = 1.0   K = 1.0

Standard elbows

90°   45°

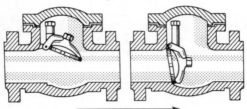

Fig. 6-4   Equivalent length and K-values for several valves and fittings

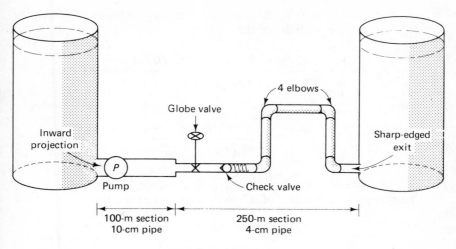

Fig. 6-5 Example 6-2

EXAMPLE 6-2

Water flows horizontally at 400 l/min (106 gpm) through commercial steel pipe in the system shown in Fig. 6-5. If the kinematic viscosity is given as 1×10^{-6} m²/s (10^{-5} ft²/sec), compute the head and power loss associated with the friction in the system.

SOLUTION

Since $z_1 = z_2$, $h_1 = h_2$, $v_1 = v_2$, and the density of the fluid does not change, the Bernoulli equation for the system is

$$h_{\text{added}} = h_{\text{losses}}$$

where

$$
\begin{aligned}
h_{\text{losses}} = \ & h_{\text{entrance}} \\
& + h_{\text{10-cm section}} \\
& + h_{\text{reduction}} \\
& + h_{\text{globe valve}} \\
& + h_{\text{check valve}} \\
& + h_{\text{4-cm section}} \\
& + h_{\text{elbows}} \\
& + h_{\text{exit}}
\end{aligned}
$$

From the continuity equation the respective velocities through the 10-cm and 4-cm sections of the pipe are

$$v_{10} = \frac{Q}{A_{10}} = \frac{(4)(400 \times 10^{-3} \text{ m}^3/\text{min})(1/60 \text{ min/s})}{(3.14)(10 \times 10^{-2} \text{ m})^2} = 0.85 \text{ m/s}$$

and

$$v_4 = \frac{Q}{A_4} = \frac{(4)(400 \times 10^{-3} \text{ m}^3/\text{min})(1/60 \text{ min/s})}{(3.14)(4 \times 10^{-2} \text{ m})^2} = 5.31 \text{ m/s}$$

The f-factors for the respective pipe sections are computed from the Reynolds numbers and relative roughness of the pipes.

$$N_{R10} = \frac{v_{10} D_{10}}{\nu} = \frac{(0.85 \text{ m/s})(10 \times 10^{-2} \text{ m})}{(10^{-6} \text{ m}^2/\text{s})} = 85\,000$$

$$N_{R4} = \frac{v_4 D_4}{\nu} = \frac{(5.31 \text{ m/s})(4 \times 10^{-2} \text{ m})}{(10^{-6} \text{ m}^2/\text{s})} = 212\,400$$

$$\text{Relative roughness}_{10} = \frac{(0.000\,046 \text{ m})}{(0.10 \text{ m})} = 0.000\,46$$

and

$$\text{Relative roughness}_4 = \frac{(0.000\,046 \text{ m})}{(0.04 \text{ m})} = 0.001$$

from which the f-factors are $f_{10} = 0.033$ and $f_4 = 0.021$, respectively.

Summing the losses, we obtain

$$h_{\text{entrance}} = K\left(\frac{v^2}{2g}\right) = \frac{(0.78)(0.85 \text{ m/s})^2}{(2)(9.8 \text{ m/s}^2)} = 0.03 \text{ m}$$

$$+ h_{\text{10-cm pipe}} = f_{10}\left(\frac{Lv^2}{D\,2g}\right) = \frac{(0.033)(100 \text{ m})(0.85 \text{ m/s})^2}{(10 \times 10^{-2} \text{ m})(2)(9.8 \text{ m/s}^2)} = 1.22 \text{ m}$$

$$+ h_{\text{reduction}} = K\left(\frac{v^2}{2g}\right) = 0.5\left(1 - \frac{D_1^2}{D_2^2}\right)\left(\frac{v^2}{2g}\right) = (0.5)(0.84)(0.037 \text{ m}) = 0.02$$

$$+ h_{\text{globe valve}} = K\left(\frac{v^2}{2g}\right) = (10)(1.44 \text{ m}) = 14.4 \text{ m}$$

$$+ h_{ck\ valve} = K\left(\frac{v^2}{2g}\right) = (2.5)(1.44\ m) = 3.60\ m$$

$$+ h_{4\text{-}cm\ pipe} = f_4\left(\frac{Lv^2}{D2g}\right) = \frac{(0.021)(250\ m)(5.31\ m/s)^2}{(4\times10^{-2}\ m)(2)(9.8\ m/s)} = 188.81\ m$$

$$+ h_{4\ elbows} = (4)(K)\left(\frac{v^2}{2g}\right) = (4)(0.9)(1.44\ m) = 5.18\ m$$

$$+ h_{exit} = K\left(\frac{v^2}{2g}\right) = (1.0)(1.44\ m) = 1.44\ m$$

and

$$h_{total} = (0.03\ m) + (1.22\ m) + (0.02\ m) + (14.4\ m) + (3.60\ m)$$
$$+ (188.81\ m) + (5.18\ m) + (1.44\ m) = 214.7\ m\ (704.4\ ft)$$

The horsepower loss associated with $h = 214.7$ m and a standard specific weight of 9802 N/m^3 [from Fig. 6-5 and Eq. (5.8)] is

$$FHP = \frac{PQ}{448} = \frac{(214.7\ m)(9802\ N/m^3)(10^{-5}\ bars/Pa)(400\ 1/min)}{448}$$

$$= 18.8\ hp$$

6-5 C-COEFFICIENTS

Some manufacturers of fluid power components prefer to express the flow characteristics of components, particularly flow control valves, in terms of a flow coefficient C. C-coefficients are also widely used to describe friction losses associated with orifices and short tubes. These will be explained in a later section.

From Torricelli's theorem, the velocity of a free stream emitted horizontally from the base of a fluid source such as a water tank is

$$v = \sqrt{2gh}$$

where the losses due to friction are assumed to be 0. If these losses are incorporated as a coefficient of the velocity,

$$v = C\sqrt{2gh}$$

and

$$h = \frac{1}{C^2}\left(\frac{v^2}{2g}\right)$$

where the value of C is usually between 0.98 and 1.00.

From Bernoulli's equation for the same system,

$$h = \frac{v^2}{2g} + h_{ff}$$

where h is the head producing the flow and h_{ff} is the component friction loss; that is,

$$h_{ff} = K\left(\frac{v^2}{2g}\right)$$

Substituting the value of h from Torricelli's theorem in the Bernoulli equation, we obtain

$$\left(\frac{1}{C^2}\right)\left(\frac{v^2}{2g}\right) = \left(\frac{v^2}{2g}\right) + K\left(\frac{v^2}{2g}\right)$$

and

$$C = 1 + \frac{1}{\sqrt{K}}$$

or

$$K = \left(\frac{1}{C^2} - 1\right) \tag{6.6}$$

6-6 EQUIVALENT LENGTH

It is evident that the head or pressure loss through a sudden enlargement, contraction, fitting, or valve is equivalent to the loss through some length of straight pipe. For example, the loss through a globe valve may be the same as the loss through several meters of the same size straight pipe. In notation

$$h_{ff} = h_f \tag{6.7}$$

where h_{ff} is the head loss through a restriction of some configuration and h_f is the loss through an equivalent length of straight pipe. Expanding these from Eqs. (6.2) and (5.20), we obtain

$$h_{ff} = K\left(\frac{v^2}{2g}\right) = f\left(\frac{Lv^2}{D2g}\right) = h_f$$

from which

$$L = D\left(\frac{K}{f}\right) \tag{6.8}$$

where L and D are in the same units, and K and f are dimensionless.

EXAMPLE 6-3

Oil with a kinematic viscosity of 5×10^{-6} m^2/s (53.82×10^{-6} ft^2/sec) is flowing through four cast iron 90-deg elbows with an inside diameter of 5 cm (2 in.) at the rate of 1900 l/m (502 gpm). Determine the equivalent length of straight pipe.

SOLUTION

From the continuity equation

$$v = \frac{Q}{A} = \frac{(4)(1900 \times 10^{-3} \text{ m}^3/\text{min})(1/60 \text{ min}/\text{s})}{(3.14)(5 \times 10^{-2})^2} = 16 \text{ m/s}$$

The Reynolds number is computed from

$$N_R = \frac{vD\rho}{\mu}$$

but from Eq. (2.14) the relationship between absolute and kinematic viscosity is

$$\nu = \frac{\mu}{\rho}$$

and

$$N_R = \frac{vD}{\nu} = \frac{(16 \text{ m/s})(5 \times 10^{-2} \text{ m})}{(5 \times 10^{-6} \text{ m}^2/\text{s})} = 160\,000$$

and the flow is turbulent.

From Table 6-1 the absolute roughness of cast iron is 0.000 28 m, and the relative roughness is computed from Eq. (6.1) as

$$\text{Relative roughness} = \frac{\epsilon}{D} = \frac{(0.000\,26 \text{ m})}{(0.05 \text{ m})} = 0.0052$$

Substituting the Reynolds number of 160 000 and relative roughness of 0.0052 in Fig. 6-1, we read the friction factor from the margin as 0.0305.

Since the four elbows are in series, their friction factor is $4 \times 0.0305 = 0.122$.

The total K-value of 3.6 for four 90-deg elbows is derived from Fig. 6-4, where the value for one elbow is 0.9.

Finally, solving for the equivalent length, using Eq. (6.8), we obtain

$$L_e = D \left(\frac{K}{f} \right) = (5 \times 10^{-2} \text{ m}) \left(\frac{3.6}{0.122} \right) = 1.48 \text{ m } (4.9 \text{ ft})$$

6-7   ECONOMIC PIPE DIAMETER

A common problem for the technican is determining the proper pipe size when the available head, length, and required flow are given, and the Reynolds number is unavailable. In these cases minor flow losses through fittings and valves are usually negligible and may be ignored, and the solution requires making progressively more accurate estimates of the friction factor such that the available head will deliver slightly more than the required flow through one of the available stock pipe sizes, or that the required flow can be delivered with slightly less than the available head. The friction factor is estimated arbitrarily at $f = 0.025$ and then progressively refined until the final estimate is within 10 percent of the last assumed value, at which time it is considered to be within the span of accepted accuracy.

From the Darcy-Weisbach equation (5.20) for turbulent flow

$$h_f = f \left(\frac{Lv^2}{D 2g} \right)$$

But from the continuity equation

$$Q = Av$$

where

$$v = \frac{4Q}{\pi D^2}$$

so that

$$h_f = f \left(\frac{L}{D 2g} \right) \left(\frac{4Q}{\pi D^2} \right)^2$$

$$h_f = 0.81 f \left(\frac{LQ^2}{D^5 g} \right) \tag{6.9}$$

and

$$D^5 = 0.81f\left(\frac{LQ^2}{h_f g}\right) \tag{6.10}$$

An example will make the use of this expression clear and will demonstrate how the results from estimates of the friction factor are corroborated.

EXAMPLE 6-4

Determine the diameter of a concrete pipe necessary to carry water at the rate of 20 000 liters/minute (5283 gpm) from one reservoir to another located 8 km (26 246 ft) away if the friction loss is not to exceed 150 m (492 ft). Assume the kinematic viscosity of water to be 1.3×10^{-6} m²/s (14×10^{-6} ft²/sec).

SOLUTION
Assuming a friction factor $f = 0.025$, and substituting in Eq. (6.10), we obtain

$$D^5 = 0.81f\left(\frac{LQ^2}{h_f g}\right)$$

$$= \frac{(0.81)(0.025)(8 \times 10^3 \text{ m})(20\,000 \times 10^{-3} \text{ m}^3/\text{min} \times 1/60 \text{ min/s})^2}{(150\text{m})(9.8\text{m/s}^2)}$$

$$D^5 = 0.012 \text{ m}^5$$

and

$$D = 0.414 \text{ m } (1.36 \text{ ft})$$

Since the pipe diameter was computed using an assumed value of the friction factor f, this must now be corroborated by determining the fluid velocity from the continuity equation, computing the Reynolds number, solving for the relative roughness of the pipe, and then reading the friction factor from Fig. 6-1. This value must be within 10 percent of the previous estimate (0.025).

Substituting in the continuity equation and solving for the Reynolds number, we obtain

$$v = \frac{Q}{A} = \frac{(4)(20\,000 \times 10^{-3} \text{ m}^3/\text{min})(1/60 \text{ min/s})}{(3.14)(0.414 \text{ m})^2} = 2.477 \text{ m/s}$$

and

$$N_R = \frac{vD}{\nu} = \frac{(2.477 \text{ m/s})(0.414 \text{ m})}{(1.3 \times 10^{-6} \text{ m}^2/\text{s})} = 788\,979$$

Solving for the relative roughness where the average absolute rough-ness ϵ for concrete is taken from Table 6-1 as 0.001 65, we have

$$\text{Relative roughness} = \frac{\epsilon}{D} = \frac{(0.001\ 65 \text{ m})}{(0.414 \text{ m})} = 0.004$$

Substituting the Reynolds number (788 070) and relative roughness (0.004) in Fig. 6-1 and reading the friction factor f from the left margin yield a value of approximately 0.029, and the original estimate was too low. To continue with the solution to the original equation (6.9), this value is taken as the new estimate of the friction factor. That is,

$$D^5 = 0.81f\left(\frac{LQ^2}{h_f g}\right)$$

$$= \frac{(0.81)(0.029)(8 \times 10^3 \text{ m})(20\,000 \times 10^{-3} 1/\text{min} \times 1/60 \text{ min/s})^2}{(150 \text{ m})(9.8 \text{ m/s}^2)}$$

$$D^5 = 0.014 \text{ m}^5$$

and

$$D = 0.427(1.4 \text{ ft})$$

Checking the solution, we have

$$v = \frac{Q}{A} = \frac{(4)(20\,000 \times 10^{-3} \text{ m}^3/\text{min})(1/60 \text{ min/s})}{(3.14)(0.427 \text{ m})^2}$$

$$= 2.33 \text{ m/s}$$

and

$$N_R = \frac{vD}{\nu} = \frac{(2.33 \text{ m/s})(0.427 \text{ m})}{(1.3 \times 10^{-6} \text{ m}^2/\text{s})} = 765\,315$$

$$\text{Relative roughness} = \frac{\epsilon}{D} = \frac{(0.001\ 65 \text{ m})}{(0.427 \text{ m})} = 0.004$$

Again, substituting the values of the Reynolds number and relative roughness in Fig. 6-1, a friction factor of approximately 0.029 is read from the left margin, and this is within 10 percent of the previous estimate of 0.029. Thus, the solution has the necessary accuracy, and a pipe diameter of 43 cm (17 in.) would be sufficient to carry the required flow of 20 000 liters/min (5283 gpm).

While friction losses incurred pumping fluid through the line account for most of the major loss, it must be remembered that a complete accounting of the energy requires that each term in the Bernoulli equation be examined. That is,

$$z_1 + \left(\frac{p_1}{\gamma}\right) + \left(\frac{v_1^2}{2g}\right) + h_{added} = z_2 + \left(\frac{p_2}{\gamma}\right) + \left(\frac{v_2^2}{2g}\right) + h_{losses}$$

where h_{losses} results from turbulent flow through the line h_f, sudden contractions h_c, enlargements h_e, and fittings and valves h_{ff}. When the line discharges in the open, for example, the kinetic energy term $v_2^2/2g$, which is the velocity head, must be accounted for as a part of the solution.

EXAMPLE 6-5

Gasoline with a kinematic viscosity of 4×10^{-7} m²/s (4.3×10^{-6} ft²/sec) is pumped into the tank shown in Fig. 6-6 through a 10-cm (4-in.) commercial steel pipe 300 m (98.4 ft) long at 100 l/min (26.4 gpm). If minor losses through fittings and valves are ignored, what must be the pressure at the pump outlet to maintain the flow specified?

SOLUTION
The Bernoulli equation for the system is

$$0 + 0 + 0 + h_a = (20 \text{ m}) + 0 + \left(\frac{v_2^2}{2g}\right) + f\left(\frac{Lv_2^2}{D2g}\right)$$

From the continuity equation

$$v_2 = \frac{Q}{A} = \frac{(4)(1000 \times 10^{-3} \text{ m}^3/\text{min})(1/60 \text{ min/s})}{(3.14)(10 \times 10^{-2})^2} = 2.12 \text{ m/s}$$

Computing the Reynolds number, we have

$$N_R = \frac{vD}{\nu} = \frac{(2.12 \text{ m/s})(10 \times 10^{-2})}{(4 \times 10^{-7} \text{ m}^2/\text{s})} = 530\ 000$$

and the flow is turbulent.

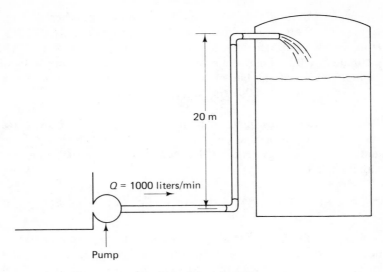

20 m

Q = 1000 liters/min

Pump

Fig. 6-6 Example 6-5

From Table 6-1, the absolute roughness of commercial steel pipe is 0.000 046 m, and the relative roughness is computed as

$$\text{Relative roughness} = \frac{(0.000\ 046\ \text{m})}{(0.10\ \text{m})} = 0.000\ 46$$

The friction factor is read from Fig. 6-1 as 0.0175, and the loss due to friction is computed from

$$h_f = f\left(\frac{Lv^2}{D2g}\right) = \frac{(0.0175)(300\ \text{m})(2.02\ \text{m/s})^2}{(10 \times 10^{-2}\ \text{m})(2)(9.8\ \text{m/s}^2)} = 10.93\ \text{m}$$

Substituting in the Bernoulli equation, we obtain

$$h_a = (20\ \text{m}) + \frac{(2.12^2)}{2g} + 10.93\ \text{m} = 31.16\ \text{m}\ (102.2\ \text{ft})$$

When the velocity is not known, the friction factor is estimated and refined through successive approximations as in Example 6-4.

6-8 DIVIDED FLOW

Divided flow is one of several pipe problems that arise from the need to provide alternate routes for fluid flow in a pipe network. City water systems, for example, provide loop systems to prevent deadheading and

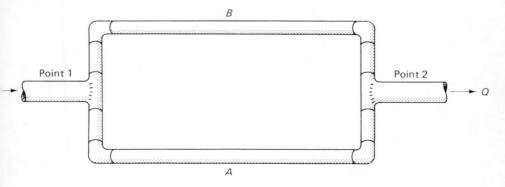

Fig. 6-7 Divided flow

stagnation, which occur when a branch circuit terminates at the end of a long run. Loop systems also provide water from more than one source to each branch during peak demands, increasing the supply, and allow isolation of leaks for repair without disrupting water service. A simple case of divided flow is illustrated in Fig. 6-7, where flow from point 1 to point 2 proceeds through routes A and B. From the figure, it is evident that the pressure drop and loss across each branch are equal, since they have a common juncture, and that the sum of the flow in branch A and branch B must equal the total flow. That is,

$$h_{fA} = f_A \left(\frac{L_A v_A^2}{D_A 2g} \right) = f_B \left(\frac{L_B v_B^2}{D_B 2g} \right) = h_{fB} \tag{6.11}$$

and

$$Q_T = Q_A + Q_B \tag{6.12}$$

or

$$\frac{\pi D^2 v}{4} = \frac{\pi D_A^2 v_A}{4} + \frac{\pi D_B^2 v_B}{4} \tag{6.13}$$

where Eqs. (6.11) and (6.12) may be solved simultaneously for v_A and v_B when the diameter and characteristics of the pipes are known. Typical of this problem is the case where a branch line divides the flow for some distance and then rejoins the main line, and it is desirable to know the flow rates in each pipe, given the pressure or flow rate at points A and B.

EXAMPLE 6-6

A 30-cm (11.8-in.) diameter water line 350 m (1148 ft) long that delivers water at 10 000 l/min (2642 gpm) branches into a 20-cm (7.9-in.) diameter line, which rejoins the main line after 700 m (2296

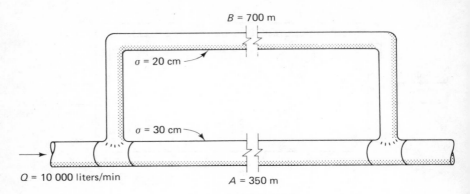

Fig. 6-8 Example 6-6

ft). Assuming a friction factor of $f = 0.025$ for each line, calculate the flow rate in the 30-cm (11.8-in.) and 20-cm (7.9-in.) diameter lines.

SOLUTION

Referring to Fig. 6-8 and substituting in Eq. (6.13), we obtain

$$Q_t = \frac{\pi D_A^2 v_A}{4} + \frac{\pi D_B^2 v_A}{4}$$

$$(10\,000 \times 10^{-3}\ \text{m}^3/\text{min} \times 1/60\ \text{min}/\text{s}) = \frac{(3.14)(30 \times 10^{-2}\ \text{m})^2(v_A)}{4}$$

$$+ \frac{(3.14)(20 \times 10^{-2}\ \text{m})^2(v_A)}{4}$$

and

$$(0.0707\ \text{m}^2)(v_A) + (0.0314\ \text{m}^2)(v_B) = 0.167\ \text{m}^3/\text{s}$$

From Eq. (6.11)

$$h_{fA} = h_{fB}$$

and

$$\frac{(0.025)(300\ \text{m})(v_A^2)}{(30 \times 10^{-2}\ \text{m})(2)(9.8\ \text{m/s}^2)} = \frac{(0.025)(700\ \text{m})(v_B^2)}{(20 \times 10^{-2}\ \text{m})(2)(9.8\ \text{m/s}^2)}$$

$$(1.28\ \text{s}^2/\text{m})(v_A^2) = (4.46\ \text{s}^2/\text{m})(v_B^2)$$

from which

$$v_A^2 = \left(\frac{4.46 \text{ s}^2/\text{m}}{1.28 \text{ s}^2/\text{m}} \right)(v_B^2) = 3.48 v_B^2$$

and

$$v_A = 1.865 v_B$$

Substituting in Eq. (6.12) and solving for v_B, we obtain

$$(0.0707 \text{ m}^2)(1.865)(v_B) + (0.0314 \text{ m}^2)(v_B) = 0.167 \text{ m}^3/\text{s}$$

$$(0.1318 \text{ m}^2)(v_B) + (0.0314 \text{ m}^2)(v_B) = 0.167 \text{ m}^3/\text{s}$$

$$(0.163 \text{ m}^2)(v_B) = 0.167 \text{ m}^3/\text{s}$$

and

$$v_B = 1.02 \text{ m/s}$$

From the continuity equation

$$Q_B = A_B v_B = \frac{(3.14)(20 \times 10^{-2} \text{ m})^2 (1.02 \text{ m/s})}{4} = 0.032 \text{ m}^3/\text{s}$$

$$= 1922 \text{ l/min (508 gpm)}$$

whereupon

$$Q_A = (10\,000 \text{ l/min}) - (1922 \text{ l/min}) = 8078 \text{ l/min (2134 gpm)}$$

Although the friction factor in example 6-6 was assumed, usually it is derived from estimates that are progressively refined, where some combination of pipe characteristics, head, and required flow are given, and the problem requires solving for system discharge or branch flow and head loss. Example 6-7 illustrates the calculations.

EXAMPLE 6-7

Compute the flow rate through each line in Example 6-6 if the 30-cm section has an absolute roughness of 0.0006 m, the 20-cm section is galvanized, and the kinematic viscosity of water is given as 1.3×10^{-6} m²/s (13.99×10^{-6} ft²/sec).

SOLUTION

The problem asks that we verify the friction factor of $f = 0.025$ for both pipes and then refine the computation of the discharge through each line. This is accomplished by examining the Reynolds number and relative roughness of the pipes and then making successive improvements in the estimate of the friction factor.

For the 30-cm diameter section of water line

$$v_A = \frac{Q_A}{A_A} = \frac{(4)(8078\times10^{-3}\ \text{m}^3/\text{min})(1/60\ \text{min}/\text{sec})}{(3.14)(30\times10^{-2}\ \text{m})^2} = 1.906\ \text{m/s}$$

$$N_{RA} = \frac{v_A D}{\nu} = \frac{(1.906\ \text{m/s})(30\times10^{-2}\ \text{m})}{(1.3\times10^{-6}\ \text{m}^2/\text{s})} = 439\ 846$$

$$\text{Relative roughness} = \frac{\epsilon}{D_A} = \frac{(0.0006\ \text{m})}{(30\times10^{-2}\ \text{m})} = 0.002$$

Substituting $N_{RA} = 439\ 846$ and relative roughness $= 0.002$ in Fig. 6-1, and reading the friction factor from the margin, we find an approximate value of $f_A = 0.024$, and the estimate is within the span of accepted accuracy.

For the 20-cm diameter section of water line

$$v_B = 1.02\ \text{m/s}$$

$$N_{RB} = \frac{v_B D_B}{\nu} = \frac{(1.02\ \text{m/s})(20\times10^{-2}\ \text{m})}{(1.3\times10^{-6}\ \text{m}^2/\text{s})} = 156\ 923$$

$$\text{Relative roughness} = \frac{\epsilon}{D_B} = \frac{(0.0002\ \text{m})}{(20\times10^{-2}\ \text{m})} = 0.001$$

Substituting $N_{RB} = 156\ 923$ and relative roughness $= 0.001$ in Fig. 6-1, and reading the friction factor from the margin, we find an approximate value of $f_B = 0.0213$, which is about 15 percent lower than the original estimate and must be recomputed with this being used as the new estimate. Substituting new estimates of $f_A = 0.0245$ and $f_B = 0.0213$ in Eq. (6.11), we have

$$h_{fA} = h_{fB}$$

$$\frac{(0.0245)(300\ \text{m})(v_A^2)}{(30\times10^{-2}\ \text{m})(2)(9.8\ \text{m/s}^2)} = \frac{(0.0213)(700\ \text{m})(v_B^2)}{(20\times10^{-2}\ \text{m})(2)(9.8\ \text{m/s}^2)}$$

$$(1.25\ \text{s}^2/\text{m})(v_A^2) = (3.80\ \text{s}^2/\text{m})(v_B^2)$$

$$v_A^2 = 3.04 v_B^2$$

and

$$v_A = 1.74 v_B$$

Substituting in Eq. (6.11) and solving for v_B, we obtain

$$(0.0707 \text{ m}^2)(1.74)(v_B) + (0.0314 \text{ m}^2)(v_B) = 0.167 \text{ m}^3/\text{s}$$

$$(0.1230 \text{ m}^2)(v_B) + (0.0314 \text{ m}^2)(v_B) = 0.167 \text{ m}^3/\text{s}$$

$$(0.154 \text{ m}^2)(v_B) = 0.167 \text{ m}^3/\text{s}$$

$$v_B = 1.08 \text{ m/s}$$

and

$$v_A = (1.748)v_B = (1.748)(1.08 \text{ m/s}) = 1.89 \text{ m/s}$$

Checking the friction factor $f_B = 0.0214$, we have

$$N_{RB} = \frac{v_B D_B}{\nu} = \frac{(1.08 \text{ m/s})(20 \times 10^{-2} \text{ m})}{(1.3 \times 10^{-6} \text{ m}^2/\text{s})} = 166\ 154$$

Again, substituting the value of $N_{RB} = 166\ 154$ and relative roughness = 0.001 in Fig 6-1, and reading the friction factor from the margin, we find an approximate value of $f_B = 0.0213$, which is satisfactory.

Finally, solving for Q_A and Q_B, we obtain

$$Q_A = A_A v_A = \frac{(3.14)(30 \times 10^{-2} \text{ m})^2(1.89 \text{ m/s})}{4} = 0.1335 \text{ m}^3/\text{s}$$

$$= 8010 \text{ l/min}$$

and

$$Q_B = A_B v_B = \frac{(3.14)(20 \times 10^{-2} \text{ m})^2(1.08 \text{ m/s})}{4} = 0.0339 \text{ m}^3/\text{s}$$

$$= 2035 \text{ l/min}$$

and

$$Q_T = Q_A + Q_B = (8010 \text{ l/min}) + (2035 \text{ l/min})$$

$$= 10\ 045 \text{ l/min} \ (2654 \text{ gpm})$$

which is in error by less than $\frac{1}{2}$ percent of the total and within the accuracy limits of Fig. 6-1.

6-9   SUMMARY AND APPLICATIONS

Resistance to flow is generated by friction within the fluid. When the flow is turbulent, the f-factor in the Darcy-Weisbach formula is related primarily to the Reynolds number and relative roughness of the wall surface.

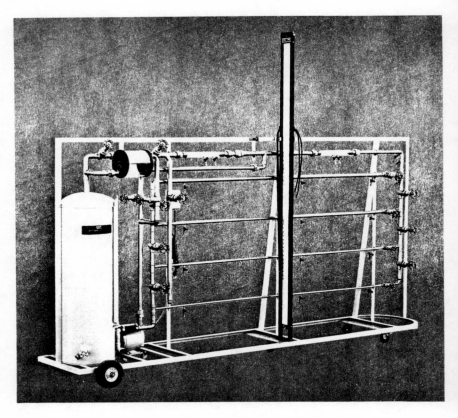

Fig. 6-9   Fluid circuit system (*Courtesy of Technovate, Inc.*)

K-values, which are experimentally determined, describe the loss through sudden enlargements, contractions, valves, and fittings as a percentage of the kinetic energy term $v^2/2g$, which is the velocity head, in the Darcy-Weisbach formula. *C*-coefficients, which are computed as percentages of the velocity term in Toricelli's theorem, are used by some manufacturers to describe the flow characteristics of fluid power components such as flow control valves. Both *K*-values and *C*-coefficients can be related to each other and used to determine equivalent lengths of straight pipe that would duplicate component losses in a system.

The utility of *f*-factors, *K*-values, and *C*-coefficients is realized in the computation of the most economic pipe diameter and component sizes, or in minimizing power losses associated with pumping a fluid between two locations. At this stage it is often necessary to assume and then refine a value for the *f*-factor such that head losses and delivery are within allowable limits for a specified pipe or component diameter and material. Divided flow poses the problem of calculating the flow rate through several branches with a common beginning and ending juncture, and requires solving simultaneous equations for two or more unknowns. In common applications such as city water systems, this would be done graphically.

Following are related applications that are useful to develop several common concepts and principles that govern resistance to fluid flow.

1. Fabricate an apparatus such as that shown in Fig. 6-9 which will demonstrate laminar and turbulent flow, and series and branch flow through pipes and fittings of different sizes.

2. Demonstrate both laminar and turbulent flow, computing Reynolds numbers associated with each through the range of operation of the system.

3. Verify K-values for sudden enlargements, sudden contractions, and several fittings and valves.

4. Verify the f-factors for four pipe sizes during completely turbulent flow.

5. Calculate the losses through a series circuit composed of pipes and fittings of varying sizes.

6. Compute the losses through a branch circuit with pipes of two diameters and different lengths.

6-10 STUDY QUESTIONS AND PROBLEMS

1. How do f-factors, K-factors, and C-coefficients differ?

2. How are f-factors and K-values related?

3. How are K-values and C-coefficients related?

4. Determine the K-value associated with a sudden enlargement that doubles the inside diameter of a pipe.

5. What would be the result if the direction of flow in Problem 4 were reversed?

6. Compute the equivalent length associated with pumping fluid through the fittings listed in Fig. 6-4 with an inside diameter of 5 cm (2 in.) if the f-factor is assumed to be 0.025.

7. What is the ratio of the equivalent length to diameter of a partially opened globe valve with a K-value of 24 and an f-factor of 0.025? The flow is considered to be completely turbulent.

8. Gasoline with a Sg of 0.68 and absolute viscosity of 2.5×10^{-4} N·s/m² (5.2×10^{-6} lbf·sec/ft²) is being pumped horizontally through a 10-cm (4-in.) diameter commercial steel pipe at 1600 l/min (423 gpm). Compute the head and horsepower loss per km of pipe associated with the friction.

9. Compute the head loss associated with pumping water horizontally at standard conditions through 150 m (492 ft) of 10-cm (4-in.) diameter cast iron pipe discharging at 1700 l/min (449 gpm).

10. A 20-cm (8-in.) diameter steel pipe carries oil with a Sg of 0.86 and an absolute viscosity of 1.85×10^{-3} N·s/m² (3.9×10^{-5} lbf·sec/ft²). What is the discharge if the head loss is not to exceed 20 m (65.6 ft) per 100 m (328 ft) of pipe?

11. Find the discharge from a one-inch inside diameter (I.D.) galvanized water pipe with a pressure drop of 4.33 psi (29 862 Pa) per 100 ft (30.5 m) at 60°F (15.6°C).

12. Compute the pipe size necessary to deliver crude oil with a kinematic viscosity of 8×10^{-5} m²/s (86×10^{-5} ft²/sec) at 4000 l/min (1057 gpm) from one reservoir to another located 2 km (6562 ft) away through a commercial steel pipe if the friction loss is not to exceed 100 m (328 ft). Assume the friction factor to be 0.025.

13. Corroborate the pipe diameter in Problem 12 by verifying the *f*-factor.

14. The loop network in Fig. 6-10 delivers water at 5000 l/min (1321 gpm). Both pipes are galvanized. If the *f*-factor is 0.025, determine the discharge through each branch.

15. If in Problem 14 the kinematic viscosity of water is 1.5×10^{-6} m²/s (16.15×10^{-6} ft²/sec), corroborate the flow rate through each pipe.

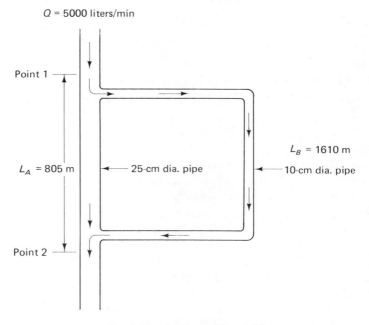

Q = 5000 liters/min

Point 1

L_A = 805 m

25-cm dia. pipe

L_B = 1610 m

10-cm dia. pipe

Point 2

Fig. 6-10   Problems 14 and 15

7

OPEN CHANNEL FLOW

7-1 INTRODUCTION

Open channel flow considers the behavior of liquids in natural and man-made channels at ambient atmospheric conditions. Rivers, canals, aqueducts, spillways, and culverts are common examples that embody related principles. In open channel flow, the upper surface of the liquid cross-section is horizontal and open to the atmosphere, and the confines of the systems are the sides and bottom of the supporting channel. Unlike hydraulic and pneumatic systems, which cause fluid to flow under the pressure generated by a pump or compressor, open channel systems rely on the potential energy due to elevation. The slope of the channel and its roughness determine the fluid velocity.

In practical situations, most open channel flow occurs in water systems at normal temperatures and atmospheric conditions, although the principles involved govern the flow of other liquids as well. Inherent in the purpose of open channel systems are efficiency of the channel cross section, controlled friction, and regulated flow rate to convey the liquid from one place to another with minimum side effects such as silting or erosion of the channel. Often, this requires dissipating some portion of the kinetic energy of the stream by directing it over surfaces of calculated roughness, weirs, and such induced phenomena as standing waves and hydraulic jumps.

Although most formulas have been empirically refined to reflect data gathered from experimentation, they have their basis in equations developed for turbulent flow in pipes. And while it is assumed that the flow is *steady* and *uniform*, i.e., that the cross section and flow characteristics, including the velocity, remain constant with respect to time and from place to place along the channel, this is rarely the case. These violations, however, do not reduce the usefulness of equations that have been developed and refined by using correcting coefficients.

7-2 HYDRAULIC RADIUS

The hydraulic radius R of an open channel is defined as the ratio of the cross-section area to the wetted perimeter in contact with the liquid. This does not include the distance across the free surface. In notation

$$\text{Hydraulic radius} = R = \frac{\text{cross-section area}}{\text{wetted perimeter}} = \frac{A}{\text{w.p.}} \qquad (7.1)$$

Hydraulic radius is sometimes referred to as the *hydraulic mean depth*.

Figure 7-1 illustrates four of the more common cross sections used in open channel flow. The round culvert in Fig. 7-1(a), for example, if it were running full would have a hydraulic radius equal to

$$R = \frac{\dfrac{\pi D^2}{4}}{\pi D} = \frac{D}{4}$$

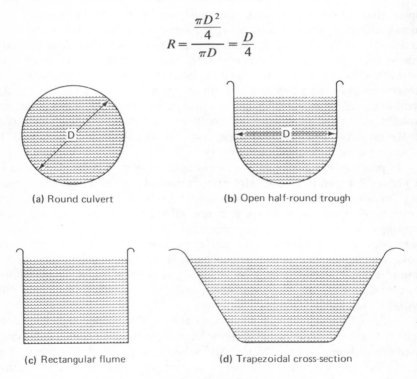

(a) Round culvert

(b) Open half-round trough

(c) Rectangular flume

(d) Trapezoidal cross-section

Fig. 7-1 Open channel cross sections

EXAMPLE 7-1

Compute the hydraulic radius for a half-round trough such as that shown in Fig. 7-1(b) if the diameter of the bottom section is 2 m (6.56 ft), the sides extend vertically 1 m (3.28 ft), and the trough is running full.

SOLUTION

Computing the hydraulic radius, we obtain

$$R = \frac{\left(\dfrac{\pi 2^2\,\mathrm{m}^2}{8}\right) + (2 \times 1\,\mathrm{m}^2)}{\left(\dfrac{\pi 2\,\mathrm{m}}{2}\right) + (2\,\mathrm{m})} = \frac{3.57\,\mathrm{m}^2}{5.14\,\mathrm{m}} = 0.69\,\mathrm{m}\ (2.26\,\mathrm{ft})$$

7-3 CHEZY AND MANNING FORMULAS

When the flow is steady and uniform, the stream has constant depth and the surface and bottom run parallel along the length of the channel. The slope of the channel S equals the ratio of the vertical fall of the stream h to the length L along the section. This is also the sine of the inclination angle θ shown in Fig. 7-2.

For slight angles of inclination, the potential energy lost as the stream flows down the inclination just equals that required to maintain the flow. Bernoulli's equation between point 1 and point 2 for the inclination is

$$z_1 + \frac{p_1}{\gamma} + \frac{v_1^2}{2g} = z_2 + \frac{p_2}{\gamma} + \frac{v_2^2}{2g} + h$$

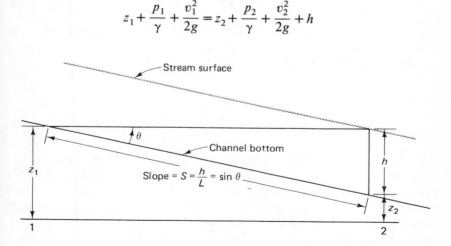

Fig. 7-2 Channel slope during steady uniform flow

and since the flow is steady and uniform, $p_1 = p_2$, $v_1 = v_2$, and

$$h = z_1 - z_2$$

From the Darcy-Weisbach equation for turbulent flow [Eq. (5.20)] for a circular cross-section channel (pipe) of length L running full,

$$h = f\left(\frac{Lv^2}{D2g}\right)$$

Replacing D with $4R$ and h/L with slope S and solving for the velocity of the stream v, we obtain

$$v^2 = \frac{2gDh}{fL}$$

$$v^2 = \left(\frac{2g}{f}\right)(4R)(S)$$

and

$$v = \sqrt{\frac{8g}{f}} \ \sqrt{RS} \tag{7.2}$$

The factor $\sqrt{8g/f}$ is usually equated to the dimensional coefficient C, and Eq. (7.2) is then written as

$$v = C\sqrt{RS} \tag{7.3}$$

where C has the dimensions $L^{1/2}T^{-1}$. This is the most recognizable form of the Chezy formula.

Since C is a function of f, that is

$$C = \sqrt{\frac{8g}{f}}$$

it is also a function of the average velocity, Reynolds number, and relative roughness of the channel surface. The effect of the kinematic viscosity of water at ordinary temperatures and inclusion of these data in actual practice is limited, because open channels are larger and rougher than most pipes. Thus, the relative roughness and Reynolds number seem to have the most pronounced effect on the C-coefficient for open channels, particularly at high Reynolds values.

Among the more widely used of the several formulas developed to account for the dimensional coefficient C is the Manning[1] equation, which states that (in English units)

$$C = \left(\frac{1.486}{n} \right)(R^{1/6}) \tag{7.4}$$

where n is a roughness factor for the channel material. If the ratio of 1.486 to the roughness factor n is replaced by the constant K, the Chezy formula has two related forms. That is,

$$v = \left(\frac{1.486}{n} \right)(R^{2/3}S^{1/2}) \tag{7.5}$$

or

$$v = KR^{2/3}S^{1/2} \tag{7.6}$$

Since C has the dimensions $L^{1/2}T^{-1}$, it follows from Eq. (7.4) that $n/1.486$ has the dimensions

$$\frac{n}{1.486} = \frac{R^{1/6}}{C} = \frac{L^{1/6}}{L^{1/2}T^{-1}} = L^{1/6}L^{-3/6}T^1 = L^{-1/3}T^1$$

In this relationship, n has the units of $ft^{1/6}$ and 1.486 has the units of $ft^{1/2}/\sec$. The use of the dimensional n and constant is awkward when more than one system of units are used, and for utility n can be assumed to be dimensionless if the constant 1.486 is given the units $L^{1/3}T^1$. In English units the constant would then have a value of 1.486 $ft^{1/3}/s$, and in SI units, it would have the value (1.486 $ft^{1/3}/s$) (0.3048 m/ft)$^{1/3}$ = 1.000 $m^{1/3}/s$.

Table 7-1 lists several of the more common materials available to construct open channels, their respective n values in English units, and K-values computed from Eq. (7.4) in SI and English units of $m^{1/3}/s$ and $ft^{1/3}/\sec$, respectively.

Turbulent flow through uniform channels of constant cross section may be computed by combining the Chezy-Manning formula and continuity equations. In SI units

$$Q = \frac{1.000AR^{2/3}S^{1/2}}{n} = \frac{AR^{2/3}S^{1/2}}{n} \text{ m}^3/\text{s} \tag{7.7}$$

where A and R have the units of m^2 and m, respectively, and in English

---

[1]Robert Manning was an Irish engineer. His original contributions describing the flow coefficient was "On The Flow Of Water In Open Channels and Pipes," *Transactions of the Institution of Civil Engineers of Ireland*, Vol. 20 (1891), p. 161, and Vol. 24 (1895), p. 179.

TABLE 7-1   *n*- and *K*-values for Chezy — Manning formula

Material surface	Relative roughness n (English units)	SI units K-value K $\dfrac{1}{(m^{3}/S)}$	English units K-value K $\dfrac{1}{(ft^{3}/S)}$
glass and plexiglass	0.010	100	149
finished concrete	0.012	83	124
glazed tile and sewer pipe	0.013	77	114
planed timber	0.013	77	114
rough timber	0.014	71	106
riveted steel	0.015	67	99
brick	0.016	63	93
stone channels	0.025	40	59
corrugated metal	0.026	38	57
gravel	0.028	36	53
earth	0.030	33	50

units

$$Q = \frac{1.486AR^{2/3}S^{1/2}}{n} \quad \text{ft}^3/\text{s} \tag{7.8}$$

where A and R have the units of ft^2 and ft, respectively. If respective K-values are substituted in Eqs. (7.7) and (7.8), the flow rate can be determined from

$$Q = KAR^{2/3}S^{1/2} \tag{7.9}$$

Common examples illustrate the use of these solutions.

EXAMPLE 7-2

If the circular conduit shown in Fig. 7-3 is 30 cm (12 in.) in diameter and is running two-thirds full, compute the hydraulic radius.

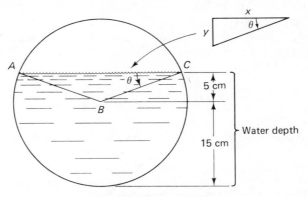

Fig. 7-3   Example 7-2

SOLUTION

At half full, the depth of the water would be 15 cm measured from the bottom of the conduit. At two-thirds full, the water level is 20 cm measured from the bottom. The angle of inclination between the radius and horizontal water surface is ACB and is computed from

$$\text{Sin}\,\theta = \frac{(5\ \text{cm})}{(15\ \text{cm})} = 0.3333$$

and $0 = 19\frac{1}{2}$ deg approximately. In triangle ABC, the central angle ABC thus equals approximately 141 deg and as a percentage this represents about 39 percent of the circumference of the circular conduit. Thus, the wetted perimeter would be

$$\text{w.p.} = 0.61\ \pi D = (0.61)(3.14)(30\ \text{cm}) = 57.46\ \text{cm}$$

The cross-section area is computed from 61 percent of the circular area plus the area in the triangular portion.[2] That is,

$$A = (0.61)\left(\frac{\pi D^2}{4}\right) + (2)\left(\tfrac{1}{2}\right)(xy)$$

$$= (0.61)(706.5\ \text{cm}^2) + (14.1\ \text{cm} \times 5\ \text{cm})$$

$$= 430.97\ \text{cm}^2 + 70.5\ \text{cm}^2 = 501.47\ \text{cm}^2$$

and the hydraulic radius is

$$R = \frac{A}{\text{w.p.}} = \frac{501.47\ \text{cm}^2}{57.46\ \text{cm}} = 8.73\ \text{cm}\ (3.4\ \text{in.})$$

EXAMPLE 7-3

Assuming that the conduit in Example 7-2 is corrugated pipe with a 0.1-m (4-in.) drop in 50 m (164 ft), compute the flow rate through the pipe.

SOLUTION

The slope equals

$$S = h/L = \frac{0.1}{50} = 0.002$$

---

[2]For convenience, the appendix lists the cross-section areas of circular pipes running partially full.

From the Chezy-Manning equation (7.7)

$$Q = \frac{AR^{2/3}S^{1/2}}{n}$$

$$= \frac{(1.000\ \text{m}^{1/3}/\text{s})(501.47 \times 10^{-4}\ \text{m}^2)(8.73 \times 10^{-2}\ \text{m})^{2/3}(0.002)^{1/2}}{(0.030)}$$

and

$$Q = (\text{m}^{1/3}/\text{s})(1.672\ \text{m}^2)(0.087\ \text{m})^{2/3}(0.002)^{1/2}$$

$$= 1.47 \times 10^{-2}\ \text{m}^3/\text{s or 882 l/min (233 gpm)}$$

EXAMPLE 7-4

A drainage ditch with a trapezoidal cross section is constructed 600 m (1969 ft) across a field with a drop of 1 m (3.28 ft). If the base of the trapezoid is 3 m (9.8 ft) and the sides slope with a pitch of 1 vertical to 3 horizontal, compute the discharge if the ditch runs with a steady uniform flow 1 m (3.28 ft) deep.

SOLUTION
With reference to Fig. 7-4, the wetted perimeter is (3 m + 6.33 m) = 9.33 and the cross-section area $A = (3\ \text{m} \times 1\ \text{m}) + (3\ \text{m} \times 1\ \text{m}) = 6\ \text{m}^2$.
Solving for R, we obtain

$$R = \frac{A}{\text{w.p.}} = \frac{(6\ \text{m}^2)}{(9.33\ \text{m})} = 0.643\ \text{m}$$

$$S = \frac{1\ \text{m}}{600\ \text{m}} = 0.0017$$

Substituting in the Chezy-Manning equation (7.9), using a K-value of 33, we have

$$Q = KAR^{2/3}S^{1/2} = (33\ \text{m}^{1/3}/\text{s})(6\ \text{m}^2)(0.643\ \text{m})^{2/3}(0.0017)^{1/2}$$

$$= 0.216\ \text{m}^3/\text{s} = 216\ \text{l/s (3424 gpm)}$$

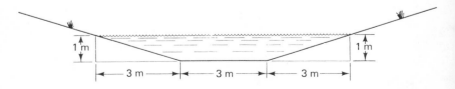

Fig. 7-4 Example 7-4

7-4 MOST EFFICIENT CROSS SECTION

The most efficient cross section of a channel is the one which for a given flow will require a minimum channel cross section and wetted perimeter. This would also result in minimum construction and material costs.

From Eq. (7.9) it is noticed that for a given velocity, the slope S, cross-section area A, and material roughness n are constant, and that maximum flow occurs through the channel when the hydraulic radius R is maximum. Since R is determined by dividing the cross-section area by the wetted perimeter, it follows that R is maximum when the wetted perimeter is minimum. Thus the channel having a minimum wetted perimeter will be the most efficient and will deliver the greatest discharge for a given cross-section area and slope. Of all regular cross-section areas, the semi-circle has the smallest wetted perimeter. For that cross section

$$R = \frac{A}{\text{w.p.}} = \frac{\frac{1}{2}(\pi r^2)}{\frac{1}{2}(2\pi r)} = \frac{r}{2} \tag{7.10}$$

and the hydraulic radius R equals half the channel radius r if the channel is running full. It follows that the most efficient configuration for any cross section is one that will accommodate a semicircle inscribed in the channel.

In determining the most efficient cross section it is common to equate the hydraulic radius R to one-half the depth of flow and then solve for base and side dimensions, slope, and flow rate.

For a trapezoid such as that shown in Fig. 7-5, the cross-section area is computed from

$$A = br + r^2 \cot\theta \tag{7.11}$$

and the wetted perimeter equals

$$\text{w.p.} = b + 2r \operatorname{cosec}\theta \tag{7.12}$$

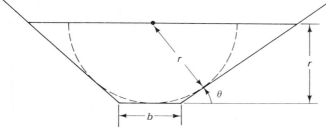

Fig. 7-5 Most efficient trapezoid channel section

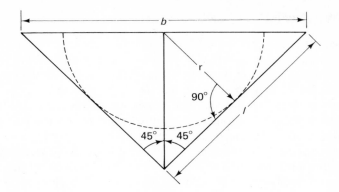

Fig. 7-6   Most efficient V-shaped channel

Since the maximum value of the hydraulic radius occurs when $R = r/2$,

$$R = \frac{A}{\text{w.p.}} = \frac{br + r^2 \cot\theta}{b + 2r \operatorname{cosec}\theta} = \frac{r}{2}$$

Solving for the base width for maximum efficiency, we obtain

$$b = 2r(\operatorname{cosec}\theta - \cot\theta) \qquad\qquad \textbf{(7.13)}$$

Or, solving for the most efficient depth, we have

$$r = \frac{b}{2(\operatorname{cosec}\theta - \cot\theta)} \qquad\qquad \textbf{(7.14)}$$

Although Eqs. (7.13) and (7.14) generalize the solution of trapezoidal cross sections, it should be noted that the most efficient of such figures that will inscribe a circle is half a hexagon with equal bottom and sides, pitched upward from the base at 60 degrees. Some materials, however, such as wet dirt, will not easily hold this shape against a flowing channel, and so Eqs. (7.13) and (7.14) provide the best available base width to side length and angle for the material available.

A rectangular channel is a special case of a trapezoid and for the most efficient cross section that will enscribe a semicircle, the channel depth is one-half the width.

For a V-shaped channel such as that in Fig. 7-6, the most efficient cross section is also the one that will inscribe a semicircle with a base angle of 45 degrees and whose radius is normal to the side at the intersection of the semicircle. Computing the hydraulic radius for such a channel, we have

$$A = (4)(\tfrac{1}{2})(r\cot\theta)(r) = 2r^2\cot\theta \qquad \textbf{(7.15)}$$

$$\text{w.p.} = 4r\cot\theta$$

and

$$R = \frac{A}{\text{w.p.}} = \frac{2r^2\cot\theta}{4r\cot\theta} = \frac{r}{2}$$

Where the central angle of the lower apex has a total angle of 90 deg, the channel depth is one-half the channel width where

$$b = 2r\operatorname{cosec}\theta = 2.828r \qquad \textbf{(7.16)}$$

and the length of the side l equals

$$l = 2r\cot\theta = 2r$$

7-5 SPECIFIC ENERGY, CRITICAL DEPTH, AND CRITICAL SLOPE

In open channel flow it is common practice to describe the energy of the flowing fluid with respect to the bed of the channel rather than some arbitrary horizontal datum line. By definition this energy is called the specific energy E and consists of the sum of the flow depth of the channel and the velocity head of the stream. That is,

$$E = y + \frac{v^2}{2g} \qquad \textbf{(7.17)}$$

where y equals the depth of the channel measured from the bottom and v is the average velocity of the stream. If channel flow rather than velocity is measured,

$$E = y + \frac{Q^2}{2gA^2} \qquad \textbf{(7.18)}$$

From Eqs. (7.17) and (7.18) the value of E, given the channel depth y, is a measure of the static pressure head. Thus for a stream of some depth, near the surface the specific energy E is primarily influenced by the kinetic energy of the stream, while near the bottom it is influenced by the static pressure.

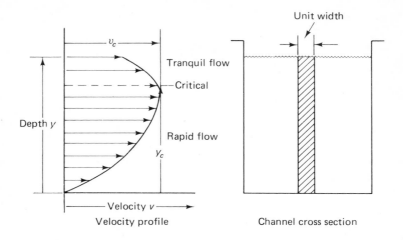

Fig. 7-7   Critical depth and velocity profile in an open rectangular channel

For a segment of the total cross-section area of unit width 1 in a rectangular channel such as that shown in Fig. 7-7, the flow rate equals

$$q = (1)(y)(v)$$

and

$$v = \frac{q}{y}$$

Substituting this value of v in Eq. (7.17), we obtain

$$E = y + \frac{q^2}{2gy^2} \qquad \textbf{(7.19)}$$

The depth where the sum of the potential energy y and kinetic energy $v^2/2g$ is minimum is termed the critical depth y_c. From theoretical analysis it can be shown that

$$y_c = \left(\frac{q^2}{g}\right)^{1/3} \quad \text{or} \quad y_c^3 = \frac{q^2}{g} \qquad \textbf{(7.20)}$$

Substituting the value of y_c in Eq. (7.19) and solving for the specific energy at the critical depth $E_{\min}$, we have

$$E_{\min} = y_c + \frac{y_c^3}{2y_c^2}$$

$$E_{\min} = \tfrac{3}{2} y_c \qquad \textbf{(7.21)}$$

and

$$y_c = \tfrac{2}{3} E_{min} \tag{7.22}$$

If the value of E_{min} is substituted in Eq. (7.17), the critical velocity at the critical depth is computed as

$$\frac{3}{2} y_c = y_c + \frac{v^2}{2g}$$

and

$$v_c = \sqrt{gy_c} \tag{7.23}$$

At depths less than the critical depth, $v < \sqrt{gy}$ and the flow is said to be rapid, supercritical, or shooting. At depths greater than the critical depth, $v > \sqrt{gy}$ and the flow is said to be slow, subcritical, or tranquil. At the critical depth, since q is assumed to be constant, the flow has the highest efficiency, since minimum energy is expended. The critical depth equals two-thirds of the specific energy, and at this depth the kinetic energy is one-half the potential energy.

Experiments to determine the velocity and flow rate of streams at varying depths are conducted using a current meter like that shown in Fig. 7-8. The apparatus consists of a rotating cup and weight suspended in the

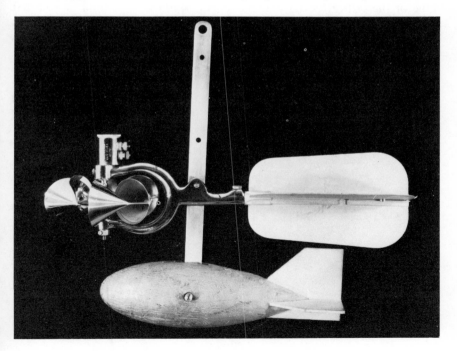

Fig. 7-8   Current meter for measuring stream velocity

stream by a line or rod. The rotating cup revolves in proportion to the velocity of the stream and transmits a signal to the surface by interrupting a set of breaker points connected to the operator's ear phones. The contacts usually interrupt once each revolution, but breaker systems interrupting every second and fifth revolution are also used. From these data, consisting of the depth at which the meter is suspended in the stream section which represents a partial volume, and the speed of rotation, the velocity and partial flow from that volume can be computed. Totaling the flow rates from the partial volumes results in the total flow rate from the stream. From the many experiments that have been conducted using this procedure it is typically found that the maximum velocity occurs at about 0.25 of the depth below the surface and the average velocity is encountered at about 0.6 of depth measured from the bottom. A more accurate method for determining the average velocity is to compute the mean of two readings taken at 0.2 and 0.8 of stream depth.

7-6　CRITICAL SLOPE

For a rectangular stream flowing uniformly in a long open channel, there is a slope that can be associated with the critical depth and velocity. This is called the critical slope S_c, which can be computed by substitution of the value of the critical velocity in the Chezy-Manning equation. Slopes that are greater than the critical slope are considered to be steep, whereas slopes less than critical are considered to be mild. At the depth that marks the division between rapid and tranquil flow there is instability, and if this depth is maintained over some distance a wavy or undulating action at the surface can be expected. It will also be noticed from Eqs. (7.20) and (7.23) that, if the depth of the stream can be made to approximate the critical depth, both the flow rate and velocity of the stream can be approximated from this single condition.

From the Chezy-Manning equation (7.9) the velocity is

$$v = KR^{2/3}S^{1/2}$$

where $K = 1/n$ in SI units and $K = 1.486/n$ in English units so that

$$v = \frac{R^{2/3}S^{1/2}}{n} \text{ m/s}$$

and

$$v = \frac{1.486R^{2/3}S^{1/2}}{n} \text{ ft/s}$$

But from Eq. (7.23) at the critical depth

$$v_c = g^{1/2} y_c^{1/2}$$

Now, by substituting the value of $v_c = Q/A$ in Eq. (7.9), it follows that for the critical slope S_c

$$v_c = g^{1/2} y_c^{1/2} = \frac{KAR^{2/3}S_c^{1/2}}{A} = KR^{2/3}S_c^{1/2} \qquad \textbf{(7.24)}$$

Solving for S_c yields

$$S_c = \frac{gy_c}{K^2 R^{4/3}} \qquad \textbf{(7.25)}$$

Substituting for K^2, in SI units, we obtain

$$S_c = \frac{n^2 gy_c}{R^{4/3}} \qquad \textbf{(7.26)}$$

and in English units

$$S_c = \frac{n^2 gy_c}{2.2 R^{4/3}} \qquad \textbf{(7.27)}$$

For wide shallow channels the value of R approaches y_c, and S_c equals

$$S_c = \frac{n^2 gy_c}{y_c^{4/3}} = \frac{n^2 g}{y_c^{1/3}} \qquad \textbf{(7.28)}$$

or in English units

$$S_c = \frac{n^2 g}{2.2 y_c^{1/3}} \qquad \textbf{(7.29)}$$

Notice that differences between the SI and English solutions for the dimensionless slope S_c are attributed to the dimensional constant and n which is in English units.

7-7 SUMMARY AND APPLICATIONS

The steady, uniform, and turbulent flow of water in open channels is influenced by the force of gravity, the slope of the channel, roughness of the wall, and the hydraulic radius or mean hydraulic depth of the channel.

Formulas that account for the velocity or flow rate from these variables or constants also incorporate a coefficient C or K to accommodate the vast amount of empirical data collected from streams, pipes, and test flumes constructed for experimental purposes.[3]

Specific energy describes the total energy of the stream with respect to the channel bottom, i.e., its flow depth and velocity head. The critical depth y_c in channels defines the depth where the specific energy is minimum, i.e., the sum of the flow depth and velocity head is minimum. This phenomenon has utility in determining the flow velocity and rate of streams using this single measurement y_c as well as in determining related characteristics of the flowing stream (rapid or tranquil flow) and channel, such as the critical slope.

The change from rapid to tranquil flow in a stream is also accompanied by a loss of energy and an increase in depth. This phenomenon is called a *hydraulic jump* or *standing wave*. The hydraulic jump is an example of nonuniform flow. While this phenomenon is undesirable in a flume or aqueduct that transports water because it reduces the efficiency of flow, it is also a valuable asset when the object is to dissipate the energy of a stream within a short distance, for example, at the toe of a dam, when it would otherwise rush down the length of the tailrace or channel, causing erosion and silting.

Following are related applications that are useful to develop several common concepts and principles that govern open channel flow. Figure 7-9 illustrates a typical flow channel for conducting these experiments. A number of other related projects are possible using actual streams, drainage ditches, and culverts available in the local community.

1. Establish several flow regimes in an open channel. such as that shown in Fig. 7-9.

2. Derive the specific energy equation from data available from an experimental flow channel.

3. Plot specific energy versus depth for a constant discharge to establish the shape of the curve on either side of the minimum energy point. To this graph, add values of E calculated directly from Eq. (7.18) for the measured value of Q.

4. Verify that the flow rate can be determined from the critical depth.

5. Verify that the critical depth will occur at the critical slope.

6. Measure the velocity profile in a stream at varying depths and locations (Fig. 7-10). Plot the velocity curve for the stream.

---

[3]The student of science history will be impressed with the material presented by Robert Manning cited in footnote 1. Manning, for example, tested his formula against 643 observations made by other experimenters studying the same phenomenon in nearly every major stream and channel in the world as well as in pipes and channels designed for experimental purposes.

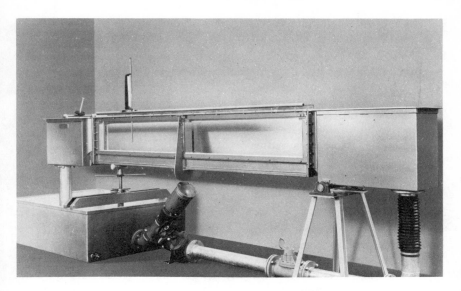

Fig. 7-9 Experimental flume (*Courtesy of Technovate, Inc.*)

Fig. 7-10 Gaging a stream (*Courtesy of Armfield Technical Education Co. Ltd. and Prof. J. Waterhouse*)

7. Demonstrate a standing wave and hydraulic jump in an experimental flow channel.

8. Calculate and verify stream flow rates, using a current meter or pitot tube and measured values of depth and velocity (Fig. 7-10).

7-8   STUDY QUESTIONS AND PROBLEMS

1. Compute the hydraulic radius of a rectangular flume 2 m (6.56 ft) wide by 2 m (6.56 ft) deep running two-thirds full.

2. Compute the hydraulic radius of a circular conduit 50 cm (19.68 in.) in diameter running three-fourths full (Fig. 7-11).

3. What are the most efficient proportions for a channel with a trapezoidal cross section?

4. A trapezoid flume made of dirt has sloping sides with a pitch such that the sides extend two meters for every one meter they rise from the horizontal plane. Determine the length of the base for the most efficient cross section.

5. What flow rate may be expected from a half-round concrete conduit with a radius of 2 m (6.56 ft) laid on a slope of 0.001 and running full?

6. What size half-round corrugated conduit having a slope of 0.002 and running full would be required to drain an area at the rate of 8.33 m^3/s (294.2 ft^3/sec)?

7. Water is flowing through a rectangular channel 3 m (9.84 ft) wide at a depth of 1 m (3.28 ft). If the flow rate is 3.5 m^3/s (123.6 ft^3/sec) and the slope is 0.001, determine the K-value for the channel material.

8. Compute the maximum velocity of a stream that can occur in a rectangular channel if the critical depth is 1.5 m (4.92 ft).

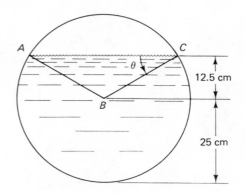

Fig. 7-11 PROBLEM 2

9. In Problem 8, compute the flow rate if the stream is 2 m (6.56 ft) wide and flows at its critical depth.

10. Calculate the specific energy for water flowing in a 1.5-m (4.92-ft) square trough running full and delivering 5 m^3/s (176.6 ft^3/sec).

11. Determine if the flow in Problem 10 is rapid or tranquil.

12. How much water will a 4 m^2 (43.1 ft^2) channel carry at a specific energy of 5 m (16.4 ft)?

13. What must be the slope to maintain a critical depth of 0.5 m (1.6 ft) in a rectangular concrete channel with a base of 2 m (6.56 ft)?

14. What slope will produce a critical velocity in a 12-m (39.4-ft) channel made of gravel running 0.15 m (0.49 ft) deep?

15. At what depth in a wide shallow channel made of earth ($n=0.030$) would a slope of 0.03 be critical?

8

IMPULSE, MOMENTUM, AND REACTION

8-1 INTRODUCTION

Flowing fluids are directed to exert forces against the members of fluid power machinery to generate movement and power. Such applications as turbines, pumps, rockets, and airplanes make use of related principles. Although the Bernoulli equation is applicable to compute the total energy available from a fluid source, it is often more convenient to account for the force generated by computing the change in momentum of the fluid. Typical are problems that direct a jet stream to impinge against moving machinery such as a turbine bucket or blade, or through a fire hose nozzle, or direct a fluid to negotiate a pipe bend that must be anchored to resist the resultant outward force. Sometimes both the Bernoulli equation and the impulse-momentum equation are used together, for example, to compute the force exerted on a pipe bend subject to both static pressure forces and the change in momentum as the fluid negotiates the curvature of the bend, as well as to changes in elevation.

In a typical high-pressure hydroelectric application, water from a high-pressure source of 300 m (984.3 ft) and above is directed against the moving member of a Pelton[1] wheel that is connected to an electric current generator.

8-2 IMPULSE-MOMENTUM EQUATION

The impulse-momentum equation is derived from Newton's second law of motion, which says in effect that the net resultant force F acting on a body is proportional to the mass M and acceleration a produced by that force.

[1]The Pelton water wheel was named after the American inventor.

162

Fig. 8-1 High-Pressure Pelton impulse wheel runner (*Courtesy of Allis-Chalmers Incorporated*)

This is written in any given direction as

$$F = Ma$$

where the acceleration equals the difference between the initial velocity v_1 and terminal velocity v_2 divided by the time t; that is,

$$a = \frac{(v_2 - v_1)}{t}$$

Notice that $(v_2 - v_1)$ is a negative quantity when mass decelerates and a positive quantity when mass accelerates, and that the force can be equated to

$$F = \frac{M}{t}(v_2 - v_1) \tag{8.1}$$

or impulse to

$$Ft = M(v_2 - v_1) \tag{8.2}$$

in which Ft is defined as the impulse, and $M(v_2 - v_1)$ is the change in momentum in a given direction. The obvious advantage that these two forms of the impulse-momentum equation have over the Bernoulli equation is that they simplify the calculation when the magnitude and direction of the resultant force is desired rather than the head loss or total energy.

From Eq. (8.1) the mass flow rate M/t can be equated to

$$\frac{M}{t} = \frac{\gamma Q}{g} = \rho Q$$

and from this relationship the component along a given direction of the impulse-momentum equation can be written in the two forms

$$F = \frac{\gamma Q}{g}(v_2 - v_1) \tag{8.3}$$

and

$$F = \rho Q(v_2 - v_1) \tag{8.4}$$

where F is the net force in newtons, γ is the specific weight of the fluid in N/m^3, ρ is the mass density of the fluid in Kg/m^3 or $N \cdot s^2/m^4$, Q is the flow rate in m^3/s, and $(v_2 - v_1)$ is the change in velocity of the fluid.

If $(v_2 - v_1)$ in Eqs. (8.3) and (8.4) is negative because the velocity is decreasing, the force F resulting from this change in velocity is opposite to the direction of flow.

8-3 MAXIMUM POWER FROM A JET

When water issuing from a round jet is directed against a flat stationary object with a large area, it can be expected to "flatten out" and disperse parallel with the surface (Fig. 8-2). From Eq. (8.3), the force exerted by the surface on the body of fluid in the opposite direction of the jet stream equals the product of the specific weight, flow rate, and change in fluid velocity divided by g. In flattening out, the direction of the jet is changed by 90 deg outward from the center of the impact area, and the velocity component in the direction of the jet is $v_2 = 0$. Thus

$$F = -\frac{\gamma Q v_1}{g}$$

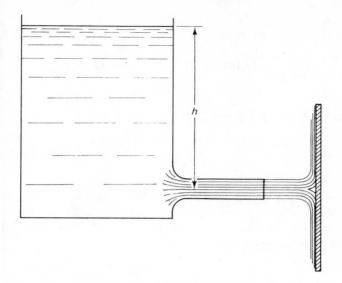

Fig. 8-2   Jet impinging on a flat plate

and

$$- F = \rho Q v_1 \qquad (8.5)$$

Where $Q = Av$,

$$- F = \frac{\gamma A v_1^2}{g} \qquad (8.6)$$

But from the Torricelli's theorem, $v^2 = 2gh$, so

$$- F = \frac{\gamma A 2gh}{g} = 2\gamma Ah = 2pA \qquad (8.7)$$

where h and p are the head and pressure on the orifice of area A. The apparent doubling of the force available from the jet when it strikes the plate results from an increase in the area acted upon by an approximate factor of six as it is redirected 90 deg and flattens out. In other words, it also indicates that h is equated to the sum of the potential and kinetic energies of the fluid as it strikes the plate, and this is greater than the potential energy alone of the fluid at rest acting against an area of the same size as the nozzle.

EXAMPLE 8-1

A jet stream 5 cm (2 in.) in diameter strikes a flat plate at right angles with a velocity of 50 m/s (164 ft/s). Compute the force exerted on the plate.

SOLUTION

From the continuity equation

$$Q = Av = \frac{(3.14)(5 \times 10^{-2} \text{ m})^2(50 \text{ m/s})}{(4)} = 0.098 \text{ m}^3/\text{s}$$

Substituting in Eq. (8.5), we obtain

$$-F = \frac{(9802 \text{ N/m}^3)(0.098 \text{ m}^3/\text{s})(50 \text{ m/s})}{(9.8 \text{ m/s}^2)} = 4.9 \text{ kN (1102 lbf)}$$

The work W in m·N available from a jet stream equals the product of the force F and distance L; the power P in watts equals the quotient W/t in joules/second; and the horsepower HP equals $P/746$. For a jet stream impinging on a flat plate that is permitted to move

$$W = FL = \frac{\gamma Q v_1 L}{g} \tag{8.8}$$

and

$$P = \frac{\gamma Q v_1 L}{gt} \tag{8.9}$$

But $\gamma L v / gt = p$, so that the power

$$P = pQ \tag{8.10}$$

Thus the power P varies with p as well as Q, but p is also a function of Q. As the flow rate is throttled closed, pressure rises until when $Q=0$, $p=h$, and $P=0$. As the flow rate is throttled open, p decreases because of the Bernoulli relationship, but the power increases because of the increase in flow rate. The power is greatest when the product pQ is maximum. It can be shown analytically that p is maximum when $p = (\frac{2}{3})h$.

EXAMPLE 8-2

A 7.5-cm (3-in.) water jet at sea level is fed from a reservoir with an elevation of 300 m (984 ft). If the jet is to be directed against the buckets of a Pelton wheel, at what pressure and flow rate would the jet deliver maximum horsepower, and what would be the efficiency?

SOLUTION

With reference to Fig. 8-2, if the maximum horsepower of the jet is realized, the pressure at the jet would be $(2/3)h$, or $h=200$ m. The velocity is computed from Torricelli's theorem

$$v = \sqrt{2gh} = \sqrt{(2)(9.8 \text{ m/s}^2)(200 \text{ m})} = 62.61 \text{ m/s}$$

From the continuity equation

$$Q = Av = \frac{(3.14)(7.5 \times 10^{-2} \text{ m})^2(62.61 \text{ m/s})}{(4)} = 0.276 \text{ m}^3/\text{s} \ (9.75 \text{ ft}^3/\text{sec})$$

The available horsepower equals

$$HP = \frac{p \times Q}{746} = \frac{(9802 \text{ N/m}^3)(200 \text{ m})(0.276 \text{ m}^3/\text{s})}{(746 \text{ J/s})} = 725 \text{ hp}$$

and the efficiency of the jet is

$$e = \frac{(200 \text{ m})}{(300 \text{ m})} \times 100 = 66.7 \text{ percent}$$

In practice, the water jet is not sized to deliver maximum horsepower, but rather is throttled to deliver jet efficiencies in the magnitude of 90 percent to conserve water.

8-4 ACTION OF A NOZZLE

Figure 8-3 illustrates the action of a nozzle. A nozzle is a converging tube section that constricts the flow from a pipe. Constricting the flow from a pipe or hose generates a set of forces in accordance with Newton's third

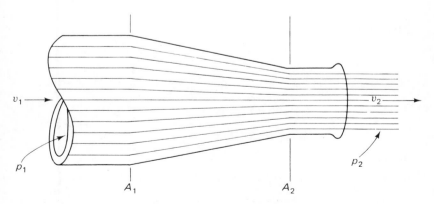

Fig. 8-3 Action of a nozzle

law of motion, which states that for every action there is an equal and opposite reaction.

At the entrance to the converging section the flow exerts a hydrostatic force in the direction of flow equal to p_1A_1. In the opposite direction at the exit of the nozzle a hydrostatic force is generated equal to p_2A_2. The force exerted by the nozzle on the fluid is also in the direction opposite the flow and is expressed as F_R. Applying the momentum impulse equation in the x direction

$$p_1A_1 + (-F_R) - p_2A_2 = \rho Q(v_2 - v_1)$$

and

$$p_1A_1 - F_R = p_2A_2 + \rho Q(v_2 - v_1) \tag{8.11}$$

Since the nozzle emits fluid to the open atmosphere, $p_2 = 0$ and

$$F_R = p_1A_1 - \rho Q(v_2 - v_1) \tag{8.12}$$

If F_R is positive, the net force on the nozzle $(-F_R)$ will be in the direction of flow, and if F_R is negative, the net force will be opposite the direction of flow. An example will illustrate the computation for related problems.

EXAMPLE 8-3

A fire hose operating at 1000 kPa (145 psi) pressure and delivering 1000 l/min (264.2 gpm) constricts the flow of water through a nozzle from an internal diameter of 4 cm (1.57 in.) to 2 cm (0.79 in.). Compute the magnitude and direction of the resultant force acting on the hose.

SOLUTION
From Eq. (8.12)

$$F_R = p_1A_1 - \rho Q(v_2 - v_1)$$

Solving for v_1 and v_2 using the continuity equation, we obtain

$$v_1 = \frac{Q}{A_1} = \frac{(4)(1000 \times 10^{-3} \text{ m}^3/\text{min})(1/60 \text{ min/s})}{(3.14)(4 \times 10^{-2} \text{ m})^2} = 13.3 \text{ m/s}$$

and

$$v_2 = \frac{Q}{A_2} = \frac{(4)(1000 \times 10^{-3} \text{ m}^3/\text{min})(1/60 \text{ min/s})}{(3.14)(2 \times 10^{-2} \text{ m})^2} = 53 \text{ m/s}$$

Substituting in Eq. (8.12) yields

$$F_R = (10 \times 10^5 \text{ N/m}^2)(1.26 \times 10^{-3} \text{ m}^2) - (1000 \text{ N/s}^2/\text{m}^4)$$
$$\cdot (1000 \times 10^{-3} \text{ m}^3/\text{min})(1/60 \text{ min/s})(53.0 \text{ m/s} - 13.3 \text{ m/s})$$
$$F_R = (1260 \text{ N}) - (662 \text{ N}) = 598 \text{ N} \text{ (134 lbf)}$$

and the resultant force on the hose is in the direction of flow. If the nozzle in Fig. 8-3, for example, were attached to an accordion-type hose, it would have a tendency to elongate and move to the right in the direction of flow.

8-5   VECTOR ANALYSIS

Fluid direction and velocity can be represented by a vector. So also can the resulting force exerted by a fluid impinging upon the members of moving machinery or flowing through bends in a pipe.

Vectors show both direction and magnitude. Direction is indicated by an arrow. Magnitude is given by the relative length of the line. Vectors are added or subtracted geometrically to determine resultant velocities and forces. By convention, vectors are summed by connecting the head end to the tail end of successive vectors and then resolving their sum or difference. The geometric sum of the two velocities (v_2) and (v_1) shown in Figs. 8-4(a) and 8-4(b) is written as

$$\text{a.} \quad (v_2) \mapsto (v_1) = v_2 \mapsto v_1 = V_1$$

which indicates that the x and y components of each velocity are added algebraically, giving the resultant vector V_1, and as

$$\text{b.} \quad (v_2) \mapsto (-v_1) = v_2 \rightarrow v_1 = V_2$$

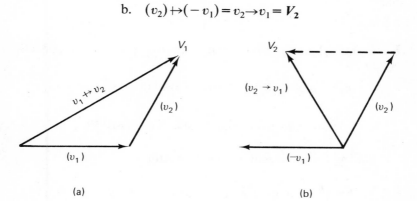

Fig. 8-4   Sum and difference of vectors

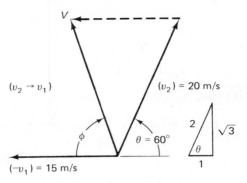

Fig. 8-5 Example 8-4

which indicates that the x and y components of (v_1) are subtracted from those of (v_2), giving the resultant V_2.

The resulting direction and magnitude of the resultant vectors can be determined by substituting the x and y components of the given velocities in the Pythagorean theorem. An example will be used to make the solution for the resultant understood.

EXAMPLE 8-4

Solve the velocity diagram shown in Fig. 8-5 for the magnitude and direction of the resultant vector.

SOLUTION

From Fig. 8-5 it is evident that

$$(v_2) \mapsto (-v_1) = v_2 \to v_1 = V$$

and

$$V = \sqrt{v_x^2 + v_y^2}$$

Resolving (v_2) and (v_1) into their respective x and y components, we obtain

$$v_x = v_2 \cos\theta - v_1 = (20 \text{ m/s})(0.5) - (15 \text{ m/s}) = -5.0 \text{ m/s}$$

and

$$v_y = v_2 \sin\theta - 0 = v_2 \sin\theta = (20 \text{ m/s})(0.866) = 17.32 \text{ m/s}$$

Solving for the magnitude of the vector V gives

$$V = \sqrt{(-5.0 \text{ m/s})^2 + (17.32 \text{ m/s})^2} = 18 \text{ m/s } (59.1 \text{ ft/s})$$

and the direction is at an angle ϕ with the horizontal and computed from

$$\cos\phi = \left(\frac{5.0 \text{ m/s}}{18 \text{ m/s}}\right) = 0.278 \quad \text{and} \quad \phi \approx 74 \text{ deg}$$

8-6   FORCE ON STATIONARY VANES

When a jet stream is redirected by a stationary vane such as that shown in Fig. 8-6, the change in fluid momentum causes a resultant force $-F_R$ to be exerted in the direction of flow, which has a tendency to move the vane in that direction. The friction generated as the fluid flows across the surface of the blade decreases the fluid velocity such that $v_2 < v_1$.

EXAMPLE 8-5

A jet stream of water delivers 0.05 m³/s (1.77 ft³/sec) against a vane with an included angle of 135 deg, at a velocity of 30 m/s (98.43 ft/s). If friction reduces the velocity to 25 m/s (82 ft/s) at the trailing edge of the vane, compute the magnitude and angle of the resultant force on the vane.

SOLUTION

Using Eq. (8.4), we compute the resultant force in the x direction as

$$F_x = \rho Q\,(v_2\cos\theta - v_1) \tag{8.13}$$

$$F_x = (1000 \text{ N}\cdot\text{s}^2/\text{m}^4)(0.05 \text{ m}^3/\text{s})\big[(25 \text{ m/s})(0.707) - (30 \text{ m/s})\big]$$

$$= -616.25 \text{ N}$$

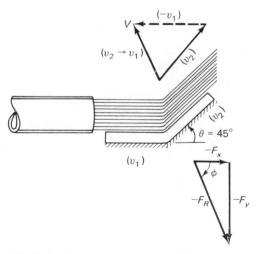

Fig. 8-6   Jet stream against a stationary vane

and because the value of F_x is negative, the horizontal force is exerted to the right in the direction of flow. In the y direction

$$F_y = \rho Q (v_2 \sin \theta - 0) \tag{8.14}$$

$$F_y = (1000 \text{ N} \cdot \text{s}^2/\text{m}^4)(0.05 \text{ m}^3/\text{s})(25 \text{ m/s})(0.707) = 883.75 \text{ N}$$

and because the value of F_y is positive, the vertical force is exerted downward. Finally, the resultant force on the vane is computed from the Pythagorean theorem

$$F_R = \sqrt{(-616 \text{ N})^2 + (884 \text{ N})^2} = 1077 \text{ N}$$

and the angle made with the horizontal plane is

$$\tan \phi = \frac{(884)}{(616)} = 1.4 \quad \text{and} \quad \theta \approx 55 \text{ deg}$$

8-7   FORCE ON MOVING VANES

So far, consideration has been given to a moving jet stream impinging on a stationary vane. The velocity of the jet was absolute with respect to the earth's surface and equaled that of the issuing jet. In actual practice, this is rarely the case. Rather, the jet, which moves with a velocity v_1, strikes the vane, which is also moving with a velocity of u. Thus the velocity of the jet that is striking the moving vane is *relative* to the vanes rather than absolute, i.e.,

$$v_{\text{relative}} = v_1 - u \tag{8.15}$$

Definite relationships exist that maximize the efficiency of vanes moving relative to the velocity of the jet which drives them.

Figure 8-7 illustrates one bucket from a Pelton wheel being driven at a velocity u by a jet with an absolute velocity v_1. The stream of water splits and is deflected 180 deg in two streams. As the water issues from the nozzle at an absolute velocity v_1, the buckets travel away from the stream at a velocity of u, and the stream enters the buckets at a velocity of $v_1 - u$. Since the stream is redirected 180 deg and exits in the opposite direction, neglecting friction (i.e., $v_2 = v_1$), the absolute exit velocity of the stream equals

$$V = (v_1 - u) - (u)$$

and

$$V = v_1 - 2u \tag{8.16}$$

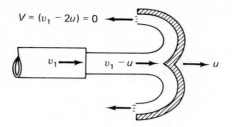

Fig. 8-7   Pelton wheel concept

If maximum efficiency were developed from the jet, then its absolute velocity V at exit from the buckets would be equal to zero.

$$v_1 - 2u = 0$$

and

$$u = \frac{v_1}{2} \tag{8.17}$$

Thus, maximum efficiency theoretically would be realized from a Pelton wheel redirecting the water at 180 deg when the velocity of the buckets was half that of the jet stream at the nozzle. The water simply would fall away free from the vanes, its velocity entirely spent. In practice, the stream is redirected less than 180 deg, so the water will clear succeeding vanes as the Pelton wheel turns, and maximum efficiencies occur when ($u = 0.43v_1$) or at about 85 percent of the velocity head of the jet.

At angles less than 180 deg, depending upon the velocities of the jet stream and the bucket vane, the redirected water stream becomes a vector quantity itself with an angle different than that of the bucket or the initial jet stream. Figure 8-8 shows the action of a jet of water impinging on a Pelton wheel, which directs the stream at an angle θ of less than 180 deg. The bucket vanes are moving away from the jet stream. Efficiency is increased by relieving the top of each vane to permit the stream to pass preceding vanes and impinge smoothly on those most normal to flow. Because the Pelton wheel rotates, bringing succeeding vanes smoothly in line with the jet, the useful flow rate is approximately that of the total stream.

For each vane, the force exerted in the x and y directions is derived from Eq. (8.4). With reference to Fig. 8-8, in the x direction the velocity equals $(v_1)_x = v_1$ the absolute velocity of the jet, and the velocity (v_2) equals

$$(v_2)_x = u + (v_1 - u)\cos\theta$$

and

$$(v_2)_x = u + (v_1 \cos\theta - u\cos\theta)$$

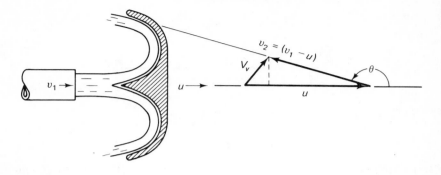

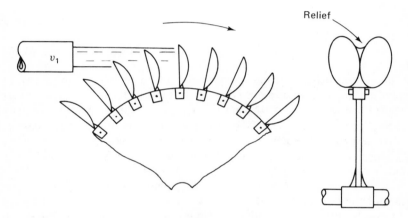

Fig. 8-8 Pelton wheel geometry

Substituting in Eq. (8.4) yields

$$F_x = \rho Q \left(u + v_1 \cos\theta - u\cos\theta - v_1 \right)$$

Rearranging terms and factoring, we have

$$F_x = \rho Q \left(\cos\theta - 1 \right)\left(v_1 - u \right) \tag{8.18}$$

In the y direction, the velocity $(v_1)_y = 0$, and the velocity $(v_2)_y$ equals

$$(v_2)_y = (v_1 - u)\sin\theta$$

Substituting in Eq. (8.4),

$$F_y = \rho Q \left(v_1 - u \right)\sin\theta \tag{8.19}$$

It should be noticed that for angles of $0° < \theta < 90°$, $\cos\theta$ will be positive, and for angles of $90° < \theta < 180°$, $\cos\theta$ will be negative. At 90 deg, $\cos\theta = 0$, $\sin\theta = 1$ and Eq. (8.18) becomes

$$F_x = \rho Q (u - v_1) \qquad (8.20)$$

which because $v_1 > u$, is a negative value and the jet imparts a positive force on the vane in the direction of the jet stream. In like manner, Eq. (8.19) becomes

$$F_y = \rho Q (v_1 - u) \qquad (8.21)$$

which is a positive value and negative force (downward in the y direction). Notice that only the force in the x direction can be converted to useful purposes and that by splitting the stream equally with the bucket, the thrust in the y direction can be cancelled. If it were not cancelled, it would become an end thrust on the supporting bearing of the Pelton wheel shaft.

The power developed by the rotating wheel is derived from the product of the force $-F_x$ times the velocity u of the vane. That is,

$$-F_x = \rho Q(v_1 - u)(1 - \cos\theta) \qquad (8.22)$$

$$\text{Power} = P = -F_x u$$

and

$$\text{Horsepower} = \frac{P}{746} = \frac{-F_x u}{746} = \frac{\rho Q u}{746}(v_1 - u)(1 - \cos\theta) \qquad (8.23)$$

where the sign is changed from Eq. (8.18) because the force on the bucket $-F_x$ is now being considered.[2]

EXAMPLE 8-6

Water issues from a 7-cm (2.76-in.) jet at 40 m/s (131.23 ft/sec) and strikes the split buckets of a rotating Pelton wheel traveling at a velocity of 25 m/s (82 ft/sec). If the curvature of the vanes is 170 deg, resolve the force of the jet on the vanes and compute the horsepower that might be developed if friction were neglected.

SOLUTION

With reference to Fig. 8-8, the force in the direction of the jet $-F_x$ is computed from Eq. (8.22)

$$-F_x = \rho Q (v_1 - u)(1 - \cos\theta)$$

[2]The student may wish to confirm that $-F_x = \rho Q(v_1 - u)(1 - \cos\theta)$ by multiplying the factors and observing that the sign of each term is changed from those of $F_x = \rho Q (v_1 - u)(\cos\theta - 1)$ when it is multiplied out.

where the flow rate is computed from the continuity equation as

$$Q = Av = \frac{(3.14)(7 \times 10^{-2} \text{ m})^2(25 \text{ m/s})}{(4)} = 0.096 \text{ m}^3/\text{s}$$

Substituting in Eq. (8.22), we obtain

$$-F_x = (1000 \text{ N} \cdot \text{s}^2/\text{m}^4)(0.096 \text{ m}^3/\text{s})(40 \text{ m/s} - 25 \text{ m/s})(1 + 0.9848)$$

and

$$-F_x = 2858 \text{ N} \ (643 \text{ lbf})$$

Thus, from Eq. (8.23) the potential horsepower developed by the Pelton wheel is

$$\text{HP} = \frac{-F_x u}{746} = \frac{(2858 \text{ N})(25 \text{ m/s})}{(746 \text{ N} \cdot \text{m/s})} = 95.8 \text{ hp}$$

8-8   FORCE ON A PIPE BEND

When fluid negotiates a pipe bend as in Fig. 8-9, the forces against the bend and connecting joints are due to both the hydrostatic pressure as well as the change in the momentum of the fluid. The resultant force has a tendency to exert side as well as end thrusts on the connecting flanges and break the bend around its outside circumference as opposed to its inside circumference.

In the x direction, the balance of forces is computed in much the same way as for a nozzle except that the nozzle is negotiating a bend. Applying the momentum impulse equation in the x direction we have

$$p_1 A_1 + (-F_x) - p_2 A_2 \cos\theta = \rho Q(v_2 \cos\theta - v_1)$$

and

$$F_x = p_1 A_1 - p_2 A_2 \cos\theta - \rho Q(v_2 \cos\theta - v_1) \tag{8.24}$$

If the cross section remains constant and flow losses are considered to be negligible, $A_1 = A_2 = A$, $p_1 = p_2 = p$, $v_1 = v_2 = v$ and

$$F_x = pA - pA\cos\theta - \rho Qv\cos\theta + \rho Qv$$

$$F_x = pA(1 - \cos\theta) + \rho Qv(1 - \cos\theta)$$

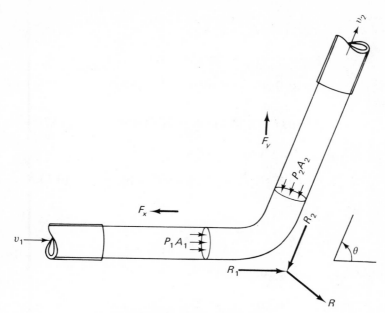

Fig. 8-9 Force on a pipe bend

and

$$F_x = (pA + \rho Qv)(1 - \cos\theta) \qquad (8.25)$$

where F_x represents the reaction of the wall of the pipe bend on the fluid opposite the direction of flow. However, this is also the force that the bend must resist in the direction of flow; thus the pipe casing and bend will be in tension.

In the y direction, if it is assumed that the pipe bend transports the fluid at the same elevation,

$$F_y = pA\sin\theta + \rho Qv\sin\theta \qquad (8.26)$$

and

$$F_y = (pA + \rho Qv)(\sin\theta) \qquad (8.27)$$

indicating that the net force will be downward on the bend in Fig. 8-9.

EXAMPLE 8-7

Water under a pressure of 15 bars (217.5 psi) is pumped through a 60-deg pipe bend like that shown in Fig. 8-9 at 4000 l/min (1057 gpm). If the inside diameter of the pipe is 10 cm (4 in.), compute the magnitude and direction of the force exerted on the bend.

SOLUTION

Solving Eq. (8.25) for F_x, we have

$$F_x = [(15 \times 10^5 \text{ N/m}^2)(78.5 \times 10^{-4} \text{ m}^2)$$

$$+ (1000 \text{ N} \cdot \text{s}^2/\text{m}^4)(0.07 \text{ m}^3/\text{s})(8.49 \text{ m/s})](1 - 0.5)$$

and

$$F_x = [(11\ 775 \text{ N}) + (566 \text{ N})](0.5) = 6170 \text{ N} = 6.2 \text{ kN}$$

Solving Eq. (8.27) for F_y, we have

$$F_y = [(11\ 775 \text{ N}) + (566 \text{ N})](0.866) = 10\ 687 \text{ N} = 10.7 \text{ kN}$$

Solving for F_R, we obtain

$$F_R = \sqrt{(6.2 \text{ kN})^2 + (10.7 \text{ kN})^2} = 12.4 \text{ kN (2780 lbf)}$$

and the direction is computed from

$$\tan\theta = \frac{10.7}{6.2} = 1.73 \quad \text{and} \quad \theta \approx 60°$$

and because F_y is positive, the direction of the angle is downward from the horizontal.

8-9 SUMMARY AND APPLICATIONS

The rush of fluid from an open jet stream generates a reaction force F_x opposite the direction of flow. If the jet stream is directed against a stationary object held normal to the stream, the resultant force is generated against the object in the direction of flow $-F_x$. Curved stationary vanes are acted upon similarly, except that the resultant force $-F_R$ is related to the x and y components of the geometry of the vane. Moving vanes introduce the velocity of the vane as an additional vector component in related problems. A nozzle increases the velocity of a jet stream.

The Pelton impulse wheel embodies the simplest application of hydraulic turbine principles. It is widely used in the generation of electric power where the head is in excess of 75 m (246 ft) for smaller units, and up to head limits that can be contained by available pipes without bursting them. Pelton wheels have been used successfully at heads above 800 m (2625 ft). Essentially, the curved buckets of the Pelton wheel convert the kinetic energy of the jet into the rotary motion of the Pelton wheel runner, as they rotate into the path of the jet to redirect and deflect the flow for maximum efficiency. Maximum power is attained when the pressure at the jet nozzle is $p=(\frac{2}{3})h$, whereas maximum efficiency occurs when the speed of the bucket equals half the velocity of the jet stream.

Following are related applications that are useful to develop several common concepts and principles that govern the momentum, impulse, and reaction caused by flowing fluids.

1. Using an air pressurized tank filled with water, corroborate Eq. (8.7) by playing a jet of water at right angles against a plate attached to a force gauge. *Note*: The surrounding area will have to be shielded from the deflected spray.

2. Construct a telescoping pipe section. Attach an adjustable nozzle constriction at the discharge end and attempt to balance the hydrostatic forces which tend to elongate the section in the direction of flow,

Fig. 8-10 Pelton wheel with dynamometer (*Courtesy of Armfield Technical Education Co., Ltd.*)

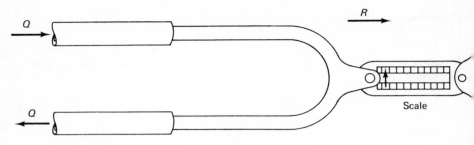

Fig. 8-11   Application 4

against the change in momentum which tends to retract the telescoping section. Compare observed data with data that might be expected
from Eq. (8.12) when $F_R = 0$.

3.   Except for size, the scale model Pelton wheel and dynamometer in Fig.
     8-10 operates in much the same way as those used for high-head
     hydroelectric applications. Using a model such as this, verify several
     operating conditions of the turbine, including jet velocity, bucket
     speed, flow rates, and horsepower output.

4.   Using a 180-deg telescoping pipe section similar to that in Fig. 8-11,
     verify the force on the scale for varying pressures and flow rates.

5.   Verify that the maximum power is available from an adjustable jet
     impinging on the runner of a Pelton wheel when $p = (\frac{2}{3})h$, and that
     maximum efficiency occurs when the velocity of the jet $v = 2u$.

6.   Construct pipe bends of 30, 45, and 90 deg and verify the resultant
     force components under various pressures and flow rates. *Note*: The
     bend must be free to move in the x and y planes.

8-10   STUDY QUESTIONS AND PROBLEMS

1.   A 2.54-cm (1-in.) diameter water jet exerts a force of 25 N (5.62 lbf) on
     a flat plate mounted to the jet. Compute the discharge of the jet.

2.   A 30-cm (11.8-in.) square water jet issues at a velocity of 1.5 m/s (4.9
     ft/sec) against a flat plate mounted normal to the jet. Compute the
     discharge of the jet and the force exerted on the flat plate.

3.   A 1.25-cm (0.5-in.) jet from a pressurized tank delivers water at 80
     l/min (21.1 gpm). What is the potential horsepower from the jet?

4.   Compute the reaction force on a 2.54-cm (1-in.) jet stream from a
     garden hose that delivers water at the rate of 200 l/min (52.83 gpm).

5.   A nozzle reduces the internal diameter of a 3.8-cm (1.5-in.) pipe
     carrying water to a jet stream 1.25 cm (0.5 in.) in diameter (Fig. 8-12).
     If the velocity in the larger section is 5 m/s (16.4 ft/sec), the hydro-

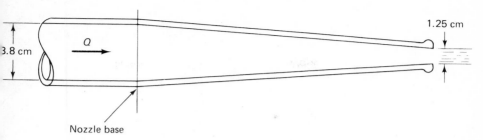

Fig. 8-12   Problem 5

static pressure at the base of the nozzle is 5500 kPa (797.5 psi), and losses through the nozzle are estimated to be 5 percent, compute the tensile or compressive force where the nozzle connects to the straight pipe.

6. Solve the vector diagram shown in Fig. 8-13 for the magnitude and direction of the resultant vector.

7. A 7.62-cm (3-in.) diameter water jet strikes a stationary curved vane at a velocity of 37 m/s (121.4 ft/sec). If the vane deflects the water through an angle of 60 deg, compute the F_x and F_y force components on the vane. Neglect friction.

8. A stationary vane redirects water delivered from a jet stream at 2000 l/min (528 gpm) through an included angle of 150 deg. If friction reduces the velocity of the jet stream from 20 m/s (65.6 ft/sec) to 15 m/s (49.2 ft/sec) at the trailing edge of the vane, compute the magnitude and angle of the resultant force on the vane.

9. Compute the maximum efficiency of a Pelton wheel from Eq. (8.22), knowing that the potential power from a free-spouting jet is ρQuv.

10. Water from a 1.25-cm (0.5-in.) diameter jet strikes the split buckets of a rotating Pelton wheel at 30 m/s (98.4 ft/sec). If the curvature of the vanes is 175 deg and friction accounts for 5 percent reduction in the velocity of the fluid as it passes through the curvature of the vanes, resolve the force of the jet on the vanes and compute the horsepower developed by the wheel. Assume an additional 5 percent mechanical loss.

11. Derive a formula for the resultant force F_r on a pipe bend if $F_x = F_y$, i.e., the bend is 90 deg.

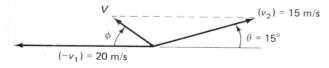

Fig. 8-13   Problem 6

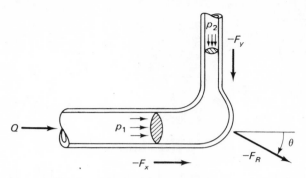

Fig. 8-14   Problem 14

12. Derive a formula for the F_x and F_y forces on a pipe bend if $p_1 \neq p_2$.

13. Derive a formula for the F_x and F_y forces on a pipe bend if $v_1 \neq v_2$, i.e., if the area of the pipe bend changes.

14. Oil with a Sg of 0.68 is redirected horizontally by a 90-deg elbow (Fig. 8.14) that reduces the internal diameter from 7.6 cm (3 in.) to 3.8 cm (1.5 in.). If the velocity and pressure in the 7.6-cm (3-in.) pipe section are 0.5 m/s (1.64 ft/sec) and 20 bars (290 psi), and the pressure loss through the elbow is given as 5 percent, compute the magnitude and direction of the resultant force on the elbow.

15. A 90-deg elbow redirects water vertically through a 7.8-cm (3.1-in.) diameter of pipe upward 10 m. If the pressure in the horizontal section is 3.5 MPa (507.5 psi) and the velocity is 7.0 m/s (23 ft/sec), compute the magnitude and direction of the resultant force on the elbow.

9

LUBRICATION MECHANICS

9-1 INTRODUCTION

Lubricants serve to reduce friction, prevent wear between moving parts, prevent adhesion between stationary parts, support moving loads, transfer heat, prevent corrosion, and even transmit power.

Lubricants consist of any material that promotes movement between parts when they are supporting a load. While petroleum base fluids, synthetic fluids, and greases are commonly associated with lubricant applications, water and air may also be used to support loads between moving bodies. Air cushion bearings, for example, are a related application.

It is common practice to fortify lubricants with additives to improve service-related properties and performance under specified conditions. The most common of these are antiwear compounds, rust and oxidation (gumming) inhibitors, foaming inhibitors, viscosity index improvers, and fire-resistant compounds.

Another characteristic of lubricants is their ability to perform under boundary conditions, that is, to sustain the lubrication function when there is a thin and unreplenished film of oil between two moving parts. Many experimental studies have been performed to explore boundary conditions.

A number of dry (solid) lubricants such as graphite, lead, babbit, phosphor-bronze, and plastic compounds are used, sometimes in conjunction with fluids, to support loads and provide primary protection against wear in high-load, low-speed applications. This occurs particularly where there are limitations to the amount of wet lubricant that can be supplied to the wear point, or where such conditions as temperature or fire hazard would adversely affect operation of the mechanism.

Filtration removes water, particulate matter, sludge, and other foreign matter from lubricants. During the process the fluid may also circulate through active material which neutralizes the acidic effects commonly associated with operation in such applications as combustion engines. In some cases, lubricants such as oils are reconditioned and replenished with additives because of replacement costs and the negative environmental effects associated with waste disposal. A number of lubricant tests have been designed to assess the service condition of fluids, either to prescribe necessary filtration, or to quantify the effects of a filtration process already in operation.

9-2   LUBRICATION MECHANICS

Dry friction occurs when two unlubricated surfaces are put in contact and move or tend to move, one relative to the other. Friction and heat are generated as surface imperfections between the two are sheared and torn away. Figure 9-1 illustrates a loaded flat plate moving parallel and in contact with a stationary surface. Where w is the load acting normal to the surface of the plate, and F is the force required to move the plate, the coefficient of friction f equals

$$f = \frac{F}{w} \tag{9.1}$$

and

$$F = fw \tag{9.2}$$

Thus, as the load or coefficient of friction increases, the force required to move the loaded plate also increases. Studies indicate that the coefficient for dry friction is relatively independent of the velocity of the moving plate, provided velocities are small and the surfaces are clean.

Viscous lubrication, as shown in Fig. 9-2, occurs when the two surfaces are separated by a film or layer of lubricant, and the shearing action and friction between the moving plate and stationary surface occurs between successive layers of the oil film separating the two. The force required to

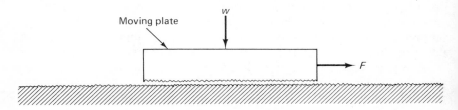

Fig. 9-1   Dry friction

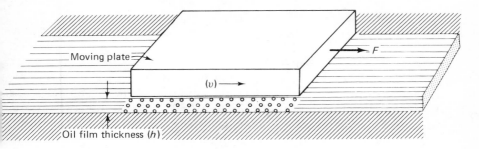

Fig. 9-2   Viscous friction

move the plate which supports a load w with a film of oil is now directly proportional ($\propto$) to the area of the plate A and shearing velocity v, and is inversely proportional to the thickness of the oil film.[1] That is,

$$F \propto \frac{Av}{h} \tag{9.3}$$

and where the proportionality constant is designated the absolute or dynamic viscosity μ

$$F = \frac{\mu Av}{h} \tag{9.4}$$

Factors in Eq. (9.4) have SI units of

$$F = \frac{\mu \ (\text{N.s/m}^2) \ A \ (\text{m}^2) \ v \ (\text{m/s})}{h \ (\text{m})} \ \text{newtons}$$

It will be noticed in Eq. (9.4) that, unlike the case of dry friction, the force required to move the plate is independent of the load, if, of course, the oil film can support it. Thus, it can be said that the heat and power loss so generated are caused by the internal friction in the oil rather than by the friction between the moving plate and stationary surface, and are independent of the load.

When a load is supported by a film of lubricant, the average pressure p developed in the oil film is computed as the ratio of the load to the area of the plate. That is,

$$p = \frac{w}{A} \tag{9.5}$$

---

[1]The present discussion parallels that in Chapter 2, Section 5, but is related more to mechanical applications than to those associated only with determining the viscosity of the fluid. It will also be noticed that the area, oil thickness, and velocity are not assigned a unit value in the notation.

and

$$w = pA \tag{9.6}$$

Notice that the oil film pressure is not the feed pressure of the lubricant; rather, the pressure is developed in the lubricant by the magnitude of the load alone. The significance of the internal oil pressure (not feed pressure) is that it has importance in determining whether or not a specific lubricant can sustain a given load. From Eqs. (9.1) and (9.6)

$$f = \frac{F}{w} = \frac{\mu A v}{hpA} = \frac{\mu v}{hp} \tag{9.7}$$

and

$$p = \frac{\mu v}{fh} \tag{9.8}$$

Thus, for a slipper bearing such as that in Fig. 9-3, the average pressure generated in the oil film is proportional to both the viscosity of the lubricant and the velocity of the plate. It is also apparent that for a given viscosity and speed, as pressure increases because of loading, the height of the oil film can be expected to decrease.

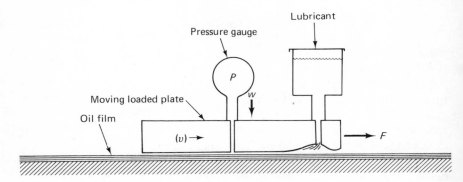

Fig. 9-3 Slipper bearing

EXAMPLE 9-1

A flat slipper bearing 20 cm × 40 cm (8 in. × 16 in.) requires a force of 250 N (56.2 lbf) to travel at a velocity of 0.5 m/s (1.6 ft/sec). If the bearing is supported by an oil film with a thickness of 7.62×10^{-5} m (0.003 in.), compute the dynamic viscosity of the lubricant.

SOLUTION

From Eq. (9.4) and referring to Fig. 9-3

$$\mu = \frac{Fh}{Av} = \frac{(250 \text{ N})(7.62 \times 10^{-5} \text{ m})}{(800 \times 10^{-4} \text{ m}^2)(0.5 \text{ m/s})}$$

$$= 4.76 \times 10^{-1} \text{ N·s/m}^2 \ (0.01 \text{ lbf·s/ft}^2)$$

EXAMPLE 9-2

If in Example 9-1 the mass supported by the bearing were 102 kg (6.99 slug), compute the pressure in the oil film and friction factor for the bearing.

SOLUTION

With reference to Fig. 9-3 and Eq. (9.5),

$$p = \frac{w}{A} = \frac{(102 \text{ kg})(9.8 \text{ m/s}^2)}{(800 \times 10^{-4} \text{ m}^2)} = 12\,495 \text{ Pa} \ (\text{N/m}^2)(1.8 \text{ psi})$$

Substituting in Eq. (9.8) and solving for f, we obtain

$$f = \frac{\mu v}{ph} = \frac{(4.76 \times 10^{-1} \text{ N·s/m}^2)(0.5 \text{ m/s})}{(12\,495 \text{ N/m}^2)(7.62 \times 10^{-5} \text{ m})} = 0.25$$

9-3 CYLINDRICAL BEARING LUBRICATION

When a loaded cylindrical shaft turns with suitable clearance in a journal fed with lubricant, the shaft floats, being supported by the viscous layer.

The action of a vertically loaded cylindrical bearing as it accelerates from a position of rest to an operating velocity is shown in Fig. 9-4. The clearance between the shaft and bearing is exaggerated for emphasis. The center of the load is indicated by the arrow; the center of the bearing by the cross marks. At rest, (a) the bearing is at the bottom of the journal, supported by a thin but continuous film of oil, preventing dry contact starting. When the shaft begins to rotate, (b) it rolls up the journal and oil that adheres to the shaft is pulled into the wedge-shaped clearance near the bottom and floats the shaft. Continued rotation (c) positions the shaft off-center opposite the lubricant wedge against which the shaft is turning and which the shaft is drawing under it by its rotation. Pressure variations are shown by the radial distribution in Fig. 9-4(c), indicating that an increase in pressure occurs starting with the converging lubricant edge and

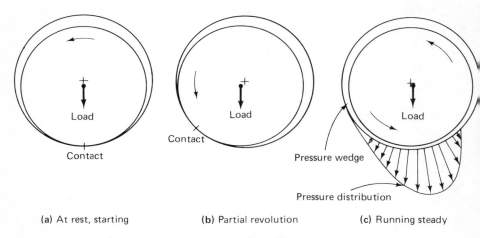

(a) At rest, starting   (b) Partial revolution   (c) Running steady

Fig. 9-4 Loaded cylindrical bearing operation

is followed by a decrease in pressure at the diverging trailing section. Thus, several statements can be made regarding similar cylindrical bearings fed with a lubricant.

1. Thin film lubrication occurs at start-up and is accompanied by the conditions for maximum bearing wear. Thus, shaft loading and quick or cold starting should be minimized during the start-up phase of shaft rotation.

2. The viscous property of the lubricant and turning of the shaft positions the vertically applied load off-center when the shaft rotates continuously.

3. Because the pressure distribution is near the bottom of the journal, as shown in Fig. 9-4(c), the lubricant should be fed near the top of the journal on the converging side of the pressure wedge, and the lower side of the journal should be kept smooth and clear of imperfections, such as notches or lubricating ports, to prevent a surface disruption that would cause a pressure drop in the oil film supporting the shaft.

4. The magnitude of the pressure developed in the lubricant is dependent upon the load rather than upon the feed pressure, which is more an indication of lubricant viscosity and shaft-to-bearing clearance than of the pressure developed by the oil film in the region supporting the rotating shaft [Fig. 9-4(c)].

5. The wedge-shaped converging oil section, which accompanies lubrication of loaded cylindrical shafts and supports the load, indicates that the viscous nature of the lubricant resists shear and "squeezing" to the extent that measurable friction and heat accompany shaft rotation.

The friction loss for cylindrical bearings is computed from a modification of Eq. (9.4). With reference to Fig. 9-5,

$$\text{Bearing area} = A = \pi D_1 L$$
$$\text{Velocity} = v = \pi D_1 N$$
$$\text{Clearance} = c = (D_2 - D_1)$$
$$\text{Lubricant thickness} = h = \frac{c}{2}$$

Substituting in Eq. (9.4), we obtain

$$F = \frac{\mu A v}{h} = \frac{(\mu)(\pi D_1 L)(\pi D_1 N)}{c/2} = \frac{2\mu\pi^2 D_1^2 NL}{c} \qquad \textbf{(9.9)}$$

From Eq. (9.7), the friction coefficient is still $f = F/w$, but unlike slipper bearings, which distribute the load w evenly across the area of the slipper, the load on cylindrical bearings is applied to the projected area of the shaft. This is illustrated by the shaded area in Fig. 9-5. That is,

$$w = pLD_1 \qquad \textbf{(9.10)}$$

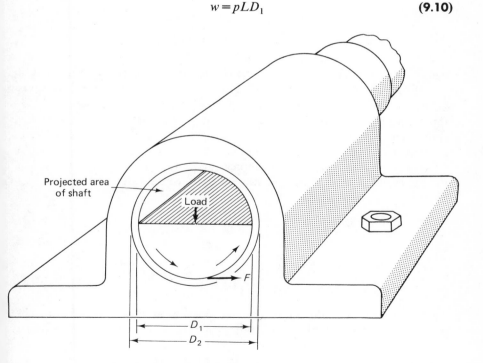

Fig. 9-5 Friction loss for cylindrical bearings

where p is the average downward force per unit of projected shaft area that must be supported by an equal and upward vertical component of oil film pressure acting against the lower half of the circumference of the shaft. Substituting Eqs. (9.9) and (9.10) into Eq. (9.7), the friction coefficient for cylindrical bearings is

$$f = \frac{F}{w} = \frac{2\mu\pi^2 D_1^2 NL}{pLD_1 c} = \frac{2\mu\pi^2 D_1 N}{pc} \qquad (9.11)$$

Notice again that in Fig. 9-4 that the lower half of the bearing supports the loaded rotating shaft off-center and on a wedge of lubricant, and that the pressure distribution shown is both off-center and asymmetrical. Because of this and the difficulty of quantifying the value of the pressure distribution, in actual practice calculations reflect the average oil film pressure derived from Eq. (9.10) rather than from conditions as they are known to exist at the circumference of the bearing.

For analysis purposes, Eq. (9.11) can be divided into its components.[2] That is,

$$f = (2\pi^2)\left(\frac{\mu N}{p}\right)\left(\frac{D_1}{c}\right) \qquad (9.12)$$

where $(2\pi^2)$, $(\mu N/p)$, and (D_1/c) are dimensionless bearing constants, and $S_2 = \mu N/p$. In practice, S_2 has a minimum value below which Eq. (9.10) is not valid. This is sometimes given as 5×10^{-8}. At values less than 5×10^{-8}, the lubricant film pressure is insufficient to support the load, with the result that boundary conditions will exist accompanied by high friction and wear. Figure 9-6 illustrates theoretical and actual values for the friction coefficient f vs $(\mu N/p)$, indicating minimum values for which Eq. (9.12) is valid. Common bearing dimensions give the value of

$$\frac{D_1}{c} = 1000 \qquad (9.13)$$

where the length of the bearing

$$L = 1.5 D_1 \qquad (9.14)$$

This considers that the lubricant will operate at a temperature of approximately 60°C (140°F). For example, a 7.64-cm (3-in.) shaft would have

---

[2]Equation (9.12) is known as the Petroff equation.

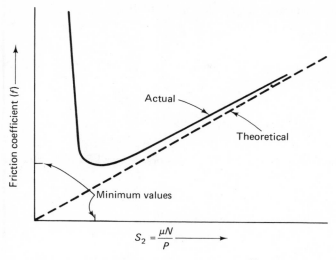

Fig. 9-6 Friction coefficient f vs $\dfrac{\mu N}{P}$

a bearing length of 11.5 cm (4.5 in.) and a clearance of

$$c = \frac{D_1}{1000} = \frac{(7.62 \times 10^{-2} \text{ m})}{(1000)} = 0.076 \text{ mm } (0.003 \text{ in.})$$

Friction power losses for cylindrical bearings are computed as the product of the tangential force to turn the shaft and the velocity v. That is,

$$Fv = \left(\frac{2\mu\pi^2 D_1^2 NL}{c} \right)(D_1 N) = \frac{2\mu\pi^3 D_1^3 N^2 L}{c} \tag{9.15}$$

in units of m·N/s. Notice that in Eq. (9.15) that friction power losses for cylindrical bearings do not depend upon the loading of the shaft.

EXAMPLE 9-2

A 5-cm (2-in.) cylindrical shaft turns in a bearing in which $L = 1.5 D_1$ and $c = D_1/1000$. If the lubricant has an absolute viscosity of 2×10^{-2} N·s/m² (4.2×10^{-4} lbf·s/ft²), compute the tangential force necessary to turn the shaft and the friction losses for the bearing.

SOLUTION
From the statement of the problem, $L = 7.5$ cm and $c = 0.005$ cm.

Substituting values in Eq. (9.9), we obtain

$$F=\frac{2\mu\pi^2 D_1^2 NL}{c}$$

$$F=\frac{(2)(2\times10^{-2}\text{ N}\cdot\text{s/m}^2)(3.14)^2(5\times10^{-2}\text{ m})^2(1500\times1/60\text{ rev/s})(7.5\times10^{-2}\text{ m})}{(0.005\times10^{-2}\text{ m})}$$

and

$$F=37\text{ N (8.3 lbf)}$$

From Eq. (9.15), the friction horsepower loss is computed as

$$\text{HP}=\frac{Fv}{746}=\frac{(37\text{ N})(3.14)(5\times10^{-2}\text{ m})(1500\times1/60\text{ rev/s})}{(746\text{ J/s})}=0.19\text{ hp}$$

EXAMPLE 9-3

From the information given in Example 9-2, compute the friction coefficient and lubricant pressure for the bearing if the shaft is supporting a mass of 90 kg.

SOLUTION

The friction factor for the bearing is given by Eq. (9.1)

$$f=\frac{F}{w}=\frac{(37\text{ N})}{(90\text{ kg})(9.8\text{ m/s}^2)}=0.042$$

and the lubricant pressure is given by Eq. (9.10)

$$p=\frac{w}{LD}=\frac{(882\text{ N})}{(7.5\times10^{-2}\text{ m})(5\times10^{-2}\text{ m})}=235\text{ kPa (34 psi)}$$

9-4   TYPES OF FLUIDS

Lubricant fluids include petroleum-base liquids, synthetic base liquids, gases, greases, and vegetable oils. Water may also be used as a lubricant, but without additives its tendency to be corrosive limits its use.

For convenience some working fluids may be classified as lubricants. For example, the oil and air used in hydraulic and pneumatic systems transmit power as a primary function, but also lubricate the machine and dissipate heat as well.

The most widely used petroleum hydraulic fluid is mineral oil, blended with additives to improve its service-related properties for specific applications and temperature ranges. Although synthetic fluids now account for less than ten percent of the lubricants and working fluids used, improved service-related properties and safety standards, particularly against fire hazards, are expected to increase their utilization. Industries and equipment that use fire-resistant fluids include forging and extrusion, coal mining, chemical-petroleum-power, die casting, foundries, fabrication, injection molding, and the primary metals industries. Less common applications include lubricating air compressors and as bearing lubricants in rotating equipment, such as steam and gas turbines, where high-temperature compressed gases can serve to promote ignition of fine oil mists present in the system. Carbon deposits, which accompany petroleum-base fluids and promote combustion, are reduced or eliminated when synthetic lubricants are used.

Grease is preferred to oil as a low-speed lubricant when the accessibility to a replenishing supply of fluid is limited, when the lubricant must also force out dirt and act as a bearing seal, and when high unit pressures and large clearances are encountered in heavy machinery applications. Non-hardening greases and improved seals have extended both the speed range and life of sealed ball bearings, used widely in such bearing applications as train axles, automotive axles, and electric motor armatures.

Synthetic fluids commonly used in hydraulic systems are of two types: straight synthetics, which are phosphate esters, and synthetic base fluids, which are compounded from phosphate esters and selected high-viscosity, low-volatility hydrocarbons.

Water glycol fluids are compounds of water (up to 50 percent by volume), glycols (one of a group of alcohols), and high-molecular-weight thickeners to increase viscosity. These fluids are compatible with hydraulic system seals used with hydrocarbon fluids, but because the water content weakens the lubricant film strength, they are limited as working fluids to 10 MPa pressure (1450 psi).

Water-in-oil emulsions are a mixture of water droplets dispersed in petroleum oil. They are also compatible with hydraulic system seals and metals (except magnesium) used with petroleum-base fluids, but have the disadvantage of separation of the emulsion. Circulation of the unemulsified water phase of the fluid will damage pumps and other components.

When air is used as a working fluid, it must be filtered to remove dirt and water, and then lubricated to protect the moving parts of the machine. Water is commonly removed by traps, by refrigerating to condense the moisture from the hot air stream leaving the compressor, or by passing through desiccants such as silica-gel, activated alumina, or molecular sieves that remove water vapors.

9-5   SERVICE-RELATED PROPERTIES

Properties that affect the performance of fluids include viscosity; specific gravity and API gravity; viscosity index; pour point; neutralization number; flash point, fire point, and auto ignition temperature; antiwear properties; resistance to oxidation and rust; and defoaming and detergent dispersant properties. Viscosity and specific gravity have been discussed previously. Selected properties of common lubricating oils are given in Table 9-1.

API degrees gravity is the second of two scales designated by the American Petroleum Industry to measure the relative weight of petroleum products. API degrees gravity and specific gravity are related by

$$\text{API degrees at } 60°F = \frac{141.5}{\text{Sg at } 60°F} - 131.5 \qquad (9.16)$$

Viscosity index is a measure of the stability of the viscosity between two temperature extremes. The viscosity of a lubricant decreases with temperature, and it becomes thinner. When the viscosity index scale was established in 1929, it measured the degree to which the viscosity of an oil would increase when it was cooled from a temperature of 210°F (98.9°C) to 100°F (37.8°C). A Pennsylvania crude paraffin base fraction was designated to have a VI of 100. A coastal crude naphtha base fraction was designated a value of 0, indicating that it changes the most with changes in temperature. Other oils were compared to these two reference oils and assigned values between 100 and 0, corresponding to their relative change in viscosity with respect to changes in temperature. Since then oils have been developed that exceed the VI of 100.

TABLE 9-1   Properties of common lubricating oils

Application	Specific Gravity	Kinematic Viscosity (m²/s @ °C)
Internal Combustion Engine Oils		
SAE 10	0.87	4.1×10^{5} @ 37.8°C, 6.0×10^{6} @ 98.9°C
SAE 30	0.89	1.1×10^{4} @ 37.8°C, 1.1×10^{5} @ 98.9°C
SAE 50	0.91	2.7×10^{4} @ 37.8°C, 2.0×10^{5} @ 98.9°C
Machine Tool Industrial Oils		
SAE 90	0.93	2.9×10^{4} @ 37.8°C, 2.1×10^{5} @ 98.9°C
SAE 140	0.93	7.3×10^{4} @ 37.8°C, 3.4×10^{5} @ 98.9°C
Turbine Oils		
Light	0.87	3.2×10^{5} @ 37.8°C, 5.5×10^{6} @ 98.9°C
Medium	0.88	6.5×10^{5} @ 37.8°C, 8.1×10^{6} @ 98.9°C
Heavy	0.89	9.9×10^{5} @ 37.8°C, 1.1×10^{5} @ 98.9°C
Hydraulic Oils		
Petroleum (mineral oil)	0.85–0.90	1.5×10^{4} @ 15°C, 2.3×10^{5} @ 49°C
Water-Glycols	1.05 avg.	4.7×10^{5} @ 37.8°C, 2.8×10^{5} @ 54°C
Synthetics (phosphate esters)	1.09–1.20	4.9×10^{5} @ 37.8°C, 5.9×10^{6} @ 98.9°C

The proper viscosity index of a fluid for a specific application is determined from the fluid temperature change requirements of the system. Hydraulic equipment, for example, such as production machinery, keeps the oil within a narrow temperature range, and an oil with a low viscosity index would be suitable. Mobile hydraulic equipment, however, operating where the temperature of the oil may vary from near 0°C to near 70°C on the outside requires a fluid with a very stable viscosity with respect to temperature change, and would call for an oil with a VI of at least 100.

Viscosity index is computed by using Saybolt Second Universal (SSU) designations for the reference oils (one with a VI of 100 and the other with a VI of 0) and for the oil for which the viscosity index is to be determined. That formula is

$$\text{VI} = \frac{(L-U)}{(L-H)} \times 100 \tag{9.17}$$

where L is the SSU viscosity of a reference oil at 100°F with a viscosity index of 0 that has the same viscosity at 210°F as the oil to be calculated; H is the SSU viscosity of a reference oil at 100°F with a viscosity index of 100 that has the same viscosity at 210°F as the oil to be calculated; and U is the viscosity of 100°F of the oil whose viscosity index is to be calculated.

In practice, the viscosity index is read directly from the tabled values in Table 9-2 by entering the SSU viscosity of the test sample determined at 210°F and 100°F. The table, composed of SSU values at 100°F, lists viscosity index values vertically down both margins and SSU values at 210°F horizontally across the top. To determine the viscosity index, the SSU viscosity values at 210°F and 100°F are located, respectively, across the top of the table and down the appropriate column, and the third value, viscosity index, is then read from the corresponding vertical margin.

EXAMPLE 9-4

Determine the viscosity index of an oil with a SSU viscosity of 105 at 210°F and 1376 at 100°F.

SOLUTION

The SSU viscosity of 90 sec at 210°F given for the oil whose viscosity index is to be determined is found by reading across the top of Table 9-2. The SSU viscosity of the sample oil at 100°F is then found by reading down the column under 105. This value is 1376. The viscosity index is then found by tracing the horizontal line of values back to the left or right margins. The VI is 95. This high value indicates a narrow range of viscosity in the temperature interval between 100–210°F.

TABLE 9-2 Viscosity index (*From Lubricating Engineers Manual, Copyright 1971, United States Steel Corporation*)

VI*

SUS 210 F	40	45	50	55	60	65	70	75	80	85	90	95	100	105	110	115	120	125	130	135	140	145	150	155	SUS 210 F
0	138	265	422	596	781	976	1182	1399	1627	1865	2115	2375	2646	2928	3220	3524	3838	4163	4498	4845	5202	5570	5959	6339	0
5	136	261	414	584	763	953	1153	1364	1585	1816	2059	2311	2573	2846	3129	3423	3727	4042	4366	4701	5046	5402	5768	6145	5
10	135	256	405	570	745	930	1124	1329	1543	1767	2002	2246	2500	2765	3038	3323	3616	3920	4233	4557	4890	5234	5587	5950	10
15	133	252	397	557	727	907	1095	1294	1502	1718	1946	2182	2427	2683	2947	3222	3505	3799	4101	4413	4734	5065	5406	5756	15
20	132	247	389	545	710	884	1066	1259	1460	1670	1889	2117	2355	2601	2856	3121	3394	3677	3968	4269	4578	4897	5225	5562	20
25	130	243	380	532	692	861	1038	1224	1418	1621	1833	2053	2282	2520	2765	3021	3284	3556	3836	4125	4423	4729	5044	5368	25
30	129	238	372	519	674	837	1009	1188	1376	1572	1776	1989	2209	2438	2674	2920	3173	3434	3703	3981	4267	4561	4863	5173	30
35	127	234	364	506	656	814	980	1153	1334	1523	1720	1924	2136	2356	2583	2819	3062	3313	3571	3837	4111	4392	4682	4979	35
40	126	230	355	493	638	791	951	1118	1293	1474	1663	1860	2063	2274	2492	2718	2951	3191	3438	3693	3955	4224	4501	4785	40
45	124	225	347	480	621	768	922	1083	1251	1425	1607	1795	1990	2193	2401	2618	2840	3070	3306	3549	3799	4056	4320	4590	45
50	123	221	339	468	603	745	893	1048	1209	1377	1551	1731	1918	2111	2311	2517	2729	2948	3173	3405	3643	3888	4139	4396	50
55	121	216	330	455	585	722	864	1013	1167	1328	1494	1667	1845	2029	2220	2416	2618	2827	3041	3261	3487	3719	3957	4202	55
60	119	212	322	442	568	699	835	978	1125	1279	1438	1602	1772	1948	2129	2316	2507	2705	2908	3117	3331	3551	3776	4007	60
65	118	207	314	429	550	676	806	943	1084	1230	1381	1538	1699	1866	2038	2215	2396	2584	2776	2937	3175	3383	3595	3813	65
70	116	203	305	416	532	653	777	908	1042	1181	1325	1473	1626	1788	1947	2114	2285	2462	2643	2829	3019	3215	3414	3619	70
75	115	194	297	403	514	630	749	873	1000	1132	1268	1409	1553	1703	1856	2014	2175	2341	2511	2685	2864	3046	3233	3425	75
80	113	190	288	391	497	606	720	837	958	1083	1212	1345	1480	1621	1765	1913	2064	2219	2378	2541	2708	2878	3052	3230	80
85	112	186	280	378	468	583	691	802	916	1035	1155	1280	1408	1539	1674	1812	1953	2098	2246	2397	2551	2710	2871	3036	85
90	110	183	272	365	461	560	662	767	875	986	1099	1216	1335	1457	1583	1711	1842	1976	2113	2253	2396	2542	2690	2842	90
95	109	181	263	352	443	537	633	732	833	937	1042	1151	1262	1376	1492	1611	1731	1855	1981	2109	2240	2373	2509	2647	95
100	107	176	255	339	426	514	604	697	791	888	986	1087	1189	1294	1401	1510	1620	1733	1848	1965	2084	2205	2328	2453	100
105	106	172	247	326	408	491	575	662	749	839	930	1023	1116	1212	1310	1409	1509	1612	1716	1821	1928	2037	2147	2259	105
110	104	167	238	314	390	468	546	627	707	790	873	958	1043	1131	1219	1309	1398	1490	1583	1677	1772	1869	1966	2064	110
115	103	163	230	301	372	445	517	592	666	741	817	894	970	1049	1128	1208	1287	1369	1451	1533	1616	1700	1785	1870	115
120	101	159	222	288	355	422	488	557	624	693	760	829	898	967	1037	1107	1176	1247	1318	1389	1460	1532	1604	1676	120
125	99	154	213	275	337	399	460	522	582	644	704	765	825	886	946	1007	1066	1126	1186	1245	1305	1364	1423	1482	125
130	98	150	205	262	319	375	431	486	540	595	647	701	752	804	855	906	955	1004	1053	1101	1149	1196	1242	1287	130
135	96	145	197	249	301	352	402	451	498	546	591	636	679	722	764	805	844	883	921	957	993	1027	1061	1093	135
140	95	141	188	236	284	329	373	416	457	497	534	572	606	640	673	704	733	761	788	813	806	859	880	899	140

Viscosity index →

*Legend

Vertical bold type—viscosity index, Horizontal bold type headings—SUS @ 210 F, tabled values are SUS @ 100° F

Comparisons of oils of different viscosity indexes can be made by plotting viscosity versus temperature curves for several samples. Figure 9-7 shows a plot of four oils with the same SSU viscosity of 90 at 100°F but with much different viscosities at 210°F. The corresponding viscosity index for each sample is listed. The procedure generates data that may be used as in Example 9-4 to determine the viscosity index.

Caution must be exercised in the interpretation and use of viscosity index at elevated temperatures. Fluids with the same SSU viscosity at 100°F and the same viscosity index may have much different viscosities at higher temperatures. That is, the relationship between viscosity and temperature may not be linear, with the result that the actual viscosity index may, in fact, be much different from the computed value outside the temperature range of 100–210°F. The implication is that viscosity index is only strictly accurate within this temperature range.

Pour point is designated as the temperature below which a fluid will not pour. Below the pour point, the oil becomes very viscous, and no movement of the surface is observable when a sample of the oil is held in a horizontal position in a test tube under set standard conditions for five seconds. The pour point is taken to be the temperature 2.78°C (5°F) above the solid point where the oil sample shows no indication of movement

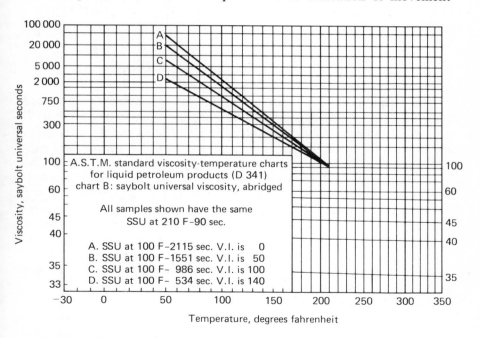

Fig. 9-7 Viscosity index for four different oils (*From* Lubrication Engineers Manual, *Copyright 1971, United States Steel Corporation*)

when the standardized test procedure adopted by the American Society for Testing and Materials (ASTM D 97) is used.

Pour point is used as an indication of the ability of lubricants, working fluids, and fuel oils to be pumped as the ambient temperature decreases. Fluids for mobile and stationary construction equipment of all types fall into this classification, and should have a pour point 10–15°C below the lowest temperature to be encountered for efficient pumping.

Neutralization number is a designation that reflects the degree of acidity or alkalinity of a lubricant. Lubricants in service have a tendency to become acid, which causes deterioration of the fluid, bearings, and other machine parts. The *acid number* refers to the number of mg of potassium hydroxide necessary to neutralize the acid in a 1-gm sample of the oil if a standardized test procedure (ASTM D 974) is used (Fig. 9-8). Neutralization is achieved by an observable color change in the solution. Conversely, the base number is the quantity of acid (alcoholic hydrochloric) in mg that is necessary to titrate the strong base constituents of a 1-gm sample of the oil. Fortified oils reduce the tendency to become acidic and keep the neutralization number below 0.1 during the normal service life of most fluids. Figure 9-8 indicates how the procedure is carried out in accordance with the ASTM designation. At the point when the color change indicates that neutralization has been reached, the neutralization number is calculated from

$$\text{Neutralization number} = \frac{\text{total ml of titrating solution} \times 5.61}{\text{weight of the sample used}} \quad \textbf{(9.18)}$$

The volatility of a lubricant in an adverse environment is described from the flash point, fire point, and auto ignition temperatures. The flash point of a fluid is the temperature at which a test flame will ignite the gases generated from a test sample that has been heated in progressive increments of 2.78°C (5°F) when it is passed near the surface of the liquid. The fire point is reached when progressive increment increases in the temperature will cause the sample to ignite and sustain a flame for five seconds when the test flame is passed near the surface. The auto ignition temperature is reached when the fluid will self-ignite and combustion is sustained continuously (ASTM Procedure D 2155). The significance of these tests indicates the magnitude of the safety hazards that are present when combustible fluids are used in the presence of hot metals, open flames, or elevated temperatures. Typical applications that present these conditions are coal mines, marine applications, aircraft, and space craft.

Antiwear properties in lubricants combat wear brought about by friction between moving parts such as shafts and bearings, piston rings and cylinder walls, and pump vanes and bodies. Wear is defined as the

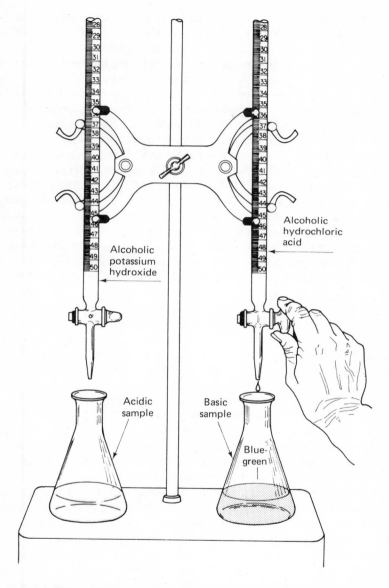

Fig. 9-8 Neutralization number test (*From* Lubrication Engineers Manual, Copyright 1971, United States Steel Corporation)

permanent displacement of surface material that occurs between the two surfaces. This causes those parts that lose material to change dimension. Typically, tests that evaluate the antiwear property of the lubricant measure the change in dimension of the parts losing material or the loss in mass (or weight) to quantify the extent of wear and effectiveness of the lubricant. When the loss is too small to measure dimensional or mass loss, parts subjected to wear may be radiated, and the loss measured as the amount of radiation present in the lubricant after the wear test.

Two standard procedures for testing the effectiveness of lubricants are commonly used: the Four-ball Wear Test (ASTM Procedure D 2266 and ASTM Procedure D 2783, Fig. 9-9), and the Vickers Hydraulic Pump Wear and Oxidation Test (ASTM Procedure D 2271 and ASTM Procedure D 2882). The Four-ball Wear Test exerts a vertical force through a rotating steel ball and measures the coefficient of friction and amount of material displaced between three stationary steel balls (or three test discs in a holding cup) and the fourth ball, which is rotated at constant speed. The contact surfaces between the test surfaces are lubricated by immersion. The relationship that defines wear-related parameters for the tests are

$$f = 2\sqrt{2}\left(\frac{F}{w}\right)\left(\frac{r}{s}\right) \tag{9.19}$$

where the coefficient of friction f is computed between 0 and 1 without dimension, the force F acting on the torque arm is measured in gm, the load L applied vertically to the rotating steel ball against the three test balls is measured in gm, the length of the torque arm r is measured in cm, and the diameter of the three stationary test balls S is measured in cm. Typical test results for the coefficient of friction with hydraulic and turbine oils range from 0.03 to 0.06, with scar diameters of 0.10 to 0.75 mm at the termination of the test, which usually lasts one hour. Scar diameters are commonly measured with a calibrated microscope.

The Vickers Pump Tests measure the wear that occurs in a vane-type pump operating as part of a hydraulic system under preset conditions of pressure, temperature, and time. Wear, viscosity, color, neutralization number, and other properties are recorded as the pump operates under the prescribed conditions for extended time intervals. Parts affected by wear (cam and vanes) and the filter screen are examined for change of weight at the termination of the test.

Oxidation produces a reaction between the lubricant and oxygen, which results in the formation of acid and sludge. Oxidation inhibitors act to reduce the catalytic effect of adjacent metals that promote the reaction, or to prevent the reaction from continuing once it has begun to occur.

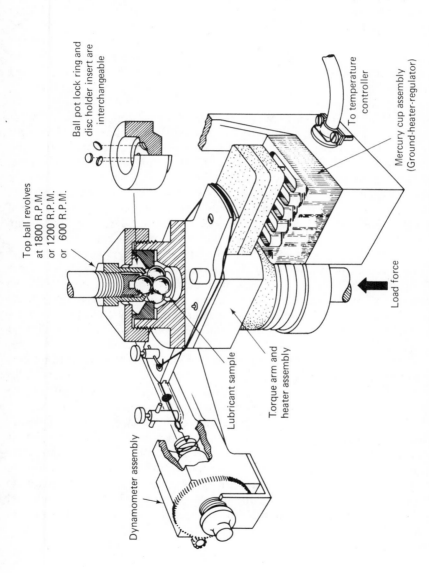

Ball pot lock ring and
disc holder insert are
interchangeable

Top ball revolves
at 1800 R.P.M.
or 1200 R.P.M.
or 600 R.P.M.

To temperature
controller

Mercury cup assembly
(Ground-heater-regulator)

Load force

Lubricant sample

Torque arm and
heater assembly

Dynamometer assembly

Fig. 9-9 Four ball wear test (*From* Lubrication Engineers Manual, *Copyright 1971, United States Steel Corporation*)

Oxidation is accelerated by elevation of the fluid temperature, particularly among the hydrocarbons, where the rate of increase doubles for every 10°C (18°F) increase. Above 60°C (140°F), it is estimated that the life of the oil is halved for every 8 + °C (15°F) rise in temperature. Tests that measure the oxidation rate and effect on lubricants (for example, ASTM D 943, ASTM D 942, ASTM D 2272, Oil Oxidation by Static Oxygen and Catalyst,[3] and Extreme Pressure Oil Oxidation: Dry Air Method[4]), accelerate the reaction by elevating the temperature to approximately 95°C in the presence of oxygen. Typically, the test is stopped when the allowed time has elapsed or the neutralization number reaches a value of 2.0.

Rusting is the corrosion of ferrous parts of machinery in the presence of water in the lubricant. Moisture enters the system in the form of condensation or through leaks in the reservoir, although exposed parts, such as greased bearings, may be subjected directly to moisture from the environment. Rusting is not to be confused with oxidation. Oxidation produces a sludge and acid condition in the lubricant, whereas rusting occurs at the metal surface itself and produces flaking. Rust inhibitors plate the ferrous surfaces, forming a thin protective coating on the metal that prevents the reaction from occurring. The standard test for rusting (ASTM D 665) subjects a polished steel rod to a mixture of 30 ml of distilled water in 300 ml of oil stirred and held at a temperature of 60°C (140°F) for 24 hours. At the termination of the test, the specimens are removed, washed in naphtha, and examined. No rust indicates Pass; the presence of rust indicates varying degrees of failure, with severe rusting being 5 percent or more of the surface coated. Figure 9-10 shows two steel rods after the rusting test, the left one in an inhibited oil and the other in a straight mineral oil.

Foaming is the result of entrainment of air in oil. Most oils contain air in solution, normally, some as much as ten percent by volume. Air in solution is not usually harmful, although it promotes oxidation of the lubricant. Air entrainment is caused by leaks near the intake or suction side of lubrication and hydraulic systems or by vigorous agitation sometimes caused by improper oil levels in the reservoir. Severe agitation is usually minimized by adding baffles inside the reservoir. Foaming depressants act to release entrained air from the fluid rather than dissolving or preventing their entrapment initially. That is, they tend to promote "breaking out" quickly so that the oil and air separate. The standard test for foaming (ASTM D 892) bubbles air through a preheated oil sample at

[3]Charles A. Bailey and Joseph S. Aarons, *The Lubrication Engineers Manual*. Cleveland: United States Steel Corporation, 1971, p. 153.
[4]*Ibid.*, p. 155.

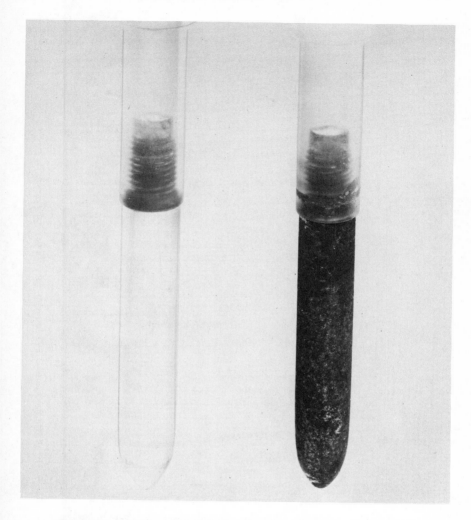

Fig. 9-10   ASTM Procedure D 665 rusting test (*Courtesy of Texaco, Inc.*)

24°C (75°F) to promote foaming and then measures the foam initially and after ten minutes of settling. After the foam has been collapsed and the sample has been cooled, the test is repeated at 93°C (200°F), and then repeated again at the initial temperature of 24°C. Figure 9-11 illustrates the effectiveness of foam depressants on hydraulic oils after 5, 10, and 15 minutes following the test and typical results for an inhibited and an uninhibited oil.

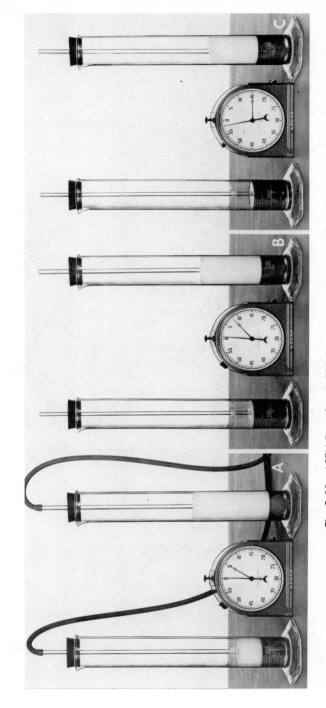

Fig. 9-11 ASTM Procedure D 892 foaming test (*Courtesy of Texaco, Inc.*)

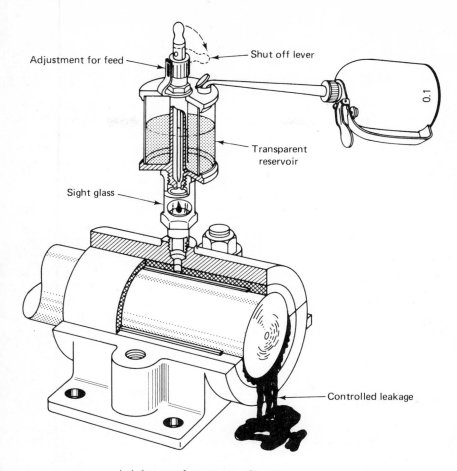

Adjustment for feed

Shut off lever

Transparent reservoir

Sight glass

Controlled leakage

0.1

Lubricant performance requirements

General requirements	Normal temperature, general purpose, straight mineral oil with correct viscosity for hand oiling, drop-feed oilers, and high-loss hydraulic systems.
Viscosity D-88	Suitable for the application intended. ASTM grades 215, 465, and 700 at 100 F suggested.
Viscosity index D-567	Not less than 50. (Low V.I. dependent upon application)
COC flast point D-92	Not less than 340 F.
Neut. No. D-974	Not more than 0.100.
Pour point D-97	No more than +20 F or lower depending on application.
Oxidation test by static oxygen	Not less than 8 hours for 60 mm pressure drop.
Four-ball wear—D2266 5 kg, 1800 rpm, 130 F, one hour	Not more than 0.6 mm.
Coefficient of friction	Not more than 0.1 under conditions of wear test.
Field test	Satisfactory performance for the application intended.

Fig. 9-12 Performance requirements for "engine oil" (*From* Lubrication Engineers Manual, *Copyright 1971, United States Steel Corporation*)

9-6   LUBRICANT PERFORMANCE REQUIREMENTS

Performance requirements consist of standards that lubricants must meet to function properly under specified conditions. They are generated in response to what are known to be normal operating conditions for a variety of machine applications. Typically, they exceed these requirements by a margin of safety to ensure that the possibility of machine damage caused by lubricant failure will be remote. Both manufacturers and users of products that have met these requirements are thus satisfied that the lubricant will provide the necessary protection against friction and other adverse conditions under which the grease or oil must operate.

Unlike standardized tests, which measure such properties as viscosity, performance requirements specify a series of tests to meet a given operating condition or set of conditions. That is, they specify which standardized tests should be conducted and passed to meet the standard. Figure 9-12 illustrates the application and lists the performance requirements for an engine oil[5] that is used in the lubrication of plain bearings, slides, machine tool bearings, and enclosed gears where extreme pressures are not encountered. In most instances, these performance requirements are standards rather than rigid specifications.[6]

9-7   FILTRATION

Strainers and filters are installed in lubricating and working circuits to remove particulate and other contaminants, such as sludge, from the fluid. Their purpose is to protect machines from sudden and catastrophic failure caused by large metallic particles migrating through the system as well as from progressive damage and failure caused from lapping by dirt and silt between the close-fitting parts of components.

Strainers are usually placed in filler tubes and pump suction lines located in the reservoir. By definition they direct the fluid in a straight line through one or more fine mesh screens, typically attached to a sheet metal support. They are cleaned by washing in solvent and blowing dry with compressed air. A typical wire mesh strainer with high flow capacity that has become loaded with contaminant is illustrated in Fig. 9-13. Four ceramic magnets permit fluid to move at low velocity through the power magnetic field. Nonferrous particles adhering to ferrous particles as small

---

[5]The term "engine oil" is a carryover from early times, when it was commonly used to identify a steam engine application, but more recently has become a designation of a general-purpose oil.
[6]A number of lubricant performance requirements are detailed in the *Lubrication Engineers Manual*, United States Steel Corporation, 1971, pp. 181–289.

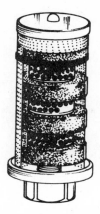

Fig. 9-13 Typical strainer equipped with ceramic magnets (*Courtesy of Schroeder Brothers Corporation*)

as 1 micron (1 micron − 0.000001 m = 0.000039 in.) are thus trapped through the viscous action of the fluid.

Filters, by definition, direct the flow of fluid through a tortuous path to extract much finer contaminants. Because they constitute a pressure loss in the filtration circuit, they are usually placed on the pressure side of the pump that circulates the oil, or in the return line to the reservoir which is at or near atmospheric pressure. Magnets and other active materials are sometimes used in their construction to increase their effect in both cleaning and neutralizing the acid in the oil. Some large plant filtration systems recondition the oil to original specifications by restoring additives lost during use and filtration.

Whereas strainers remove particles from the fluid down to about 40 microns, filters are capable of removing particles down to 1 micron. Consequently, they have a marked effect on the life of components, since by actual count more than 90 percent of the contaminants are smaller than 10 microns. The relative size of micronic particles magnified 500 times is illustrated in Fig. 9-14. Particles down to about 40 microns can be seen with the naked eye.

Filters may be part of the lubrication or working fluid circuit on the machine itself, or function as portable units that are taken throughout a plant. Portable units are used to filter lubricants from the reservoirs of several machines to remove fluid contamination accumulated during hundreds of hours of operation. Figure 9-15 illustrates a portable filtration unit consisting of a motor-driven pump and large filtration cartridge. The condition of the oil is checked after it has been filtered.

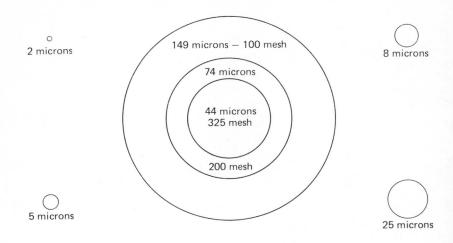

Relative sizes

Lower limit of visibility (naked eye)	40 microns
White blood cells	25 microns
Red blood cells	8 microns
Bacteria (cocci)	2 microns

Linear equivalents

1 inch	25.4 millimeters	25,400 microns
1 millimeter	0.0394 inches	1,000 microns
1 micron	25,400 of an inch	0.001 millimeters
1 micron	3.94×10^{-5}	0.000039 inches

Screen sizes

Meshes per linear inch	U.S. sieve no.	Opening in inches	Opening in microns
52.36	50	0.0117	297
72.45	70	0.0083	210
101.01	100	0.0059	149
142.86	140	0.0041	105
200.00	200	0.0029	74
270.26	270	0.0021	53
323.00	325	0.0017	44
		0.00039	10
		0.000019	.5

Fig. 9-14   Relative size of micronic particles multiplied 500 times (*Courtesy of Sperry Vickers*)

Fig. 9-15 Portable filtration unit (*Courtesy of Schroeder Brothers Corporation*)

Filter circuits locate the element in a number of places: directly in the high-pressure line, in a by-pass circuit either in the high-pressure line or return line, or in the return line where the lubricant enters the reservoir. Full-flow filters cycle all the oil each time the lubricant circulates. If the filter is located upstream of components, this has the effect of removing contaminants before the lubricant or working fluid enters the close-fitting parts of the machinery. Return and by-pass filtration circuits filter the oil separately or after it has been circulated through the system. Nearly all filters are equipped with a by-pass valve to ensure that at high flow rates, or as the filtering element becomes progressively saturated during extended use, that lubricant will continue to flow through the unit. Figure 9-16 illustrates the working parts and filter cartridge of a typical disposable full-flow filter unit.

The type of filtration circuit is influenced by system pressures, by the tolerance that the system has for contamination, by the degree of filtration desired, and by accessibility for servicing. In most instances, where the filter is located is more important than the type of circuit used, to ensure that at least a portion of the lubricant is continually circulated through a serviceable filter element. Current design practice places easily serviced (usually throwaway) elements conspicuously accessible to maintenance personnel, usually with date and service history recorded on or close by the filter for observation by the operator and supervisory personnel.

Field tests of lubricants and working fluids, commonly called *patch tests*, are conducted at the site to compare contamination counts of the test fluid under a microscope with those of reference fluids. The kit shown in Fig. 9-17 is used for all hydraulic fluids but water, glycols, and synthetics. One procedure consists of forcing the test fluid sample through a filter disc and examining it under $80\times$ magnification. The results are compared with a set of reference photos supplied with the kit. Major oil companies also provide a laboratory analysis service for oil samples taken in the field.

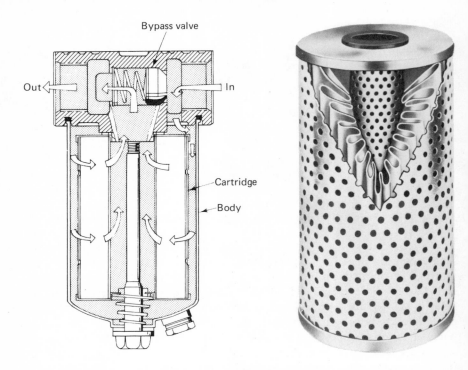

Fig. 9-16   Left—full-flow filter action with bypass (*Courtesy of Sperry Vickers*); Right—Disposable filter construction (*Courtesy of Schroeder Brothers Corporation*)

Fig. 9-17   Test kit for hydraulic fluids (*Courtesy of Gulf Oil Corporation*)

9-8   SUMMARY AND APPLICATIONS

Lubrication makes parts in contact "slippery" to support moving loads with minimum friction. This function is in addition to others that the fluid must perform, such as transmitting power, transferring heat, reducing wear, and preventing adhesion and deterioration of the fluid and machinery with which it comes in contact.

In thick film or viscous lubrication, the force required to sustain movement of the bearing surface is independent on the load, and losses so generated are caused by resistance to the relative movement of the lubricant molecules past one another. In flat bearings, the pressure generated in the oil film equals the load divided by the supporting area. In journal bearings that support the load on half the circumference, although this is not strictly so, the average oil film pressure is still taken to be the load on the shaft divided by the projected area of the shaft. In both cases, the pressure is proportional to the viscosity of the lubricant and the relative velocity of the flat plate or round journal.

The composition of the lubricant is determined by several considerations, including cost, availability, machine requirements, and environmental factors. If the lubricant is also a working fluid, such as one of the hydraulic oils, that is also an important factor in its selection. While petroleum lubricants are used as the base for the vast majority of applications, special requirements imposed by safety conditions are having the effect of increasing the use and application of synthetics and other fire-resistant fluids to protect life and insure against capital loss.

A number of properties are related to the service of lubricants, including viscosity, viscosity index, neutralization number, volatility, antiwear properties, oxidation, rusting, and foaming. In addition, filtering and testing for contaminants assures a full service life for the fluid and components. A number of standardized fluid tests and lubricant performance requirements are available to assess the condition of specific lubricants, both in the laboratory and in the field. A representative sampling of these follows. Complete specifications for each of these tests and performance requirements are available from the source cited. Whenever possible, the primary source should be used as a guide when actual tests are conducted.

1. ASTM D 91-61, reapproved 1973. Standard Method of Test for the Precipitation Number.
2. ASTM D 92-72, reapproved 1972. Standard Method of Test for Flash and Fire Points by Cleveland Open Cup Method.
3. ASTM D 97-66, reapproved 1971. Standard Method of Test for Pour Point.

4. ASTM D 217-68, reapproved 1973. Standard Method of Test for Cone Penetration of Lubricating Grease.
5. ASTM F 313-70. Standard Method of Test for Insoluble Contaminants of Hydraulic Fluids by Gravimetric Analysis.
6. ASTM F 317-72. Standard Method of Test for Liquid Flow Rate of Membrane Filters.
7. ASTM D 664-58, reapproved 1975. Standard Method of Test for Neutralization Number by Potentiometric Titration.
8. ASTM D 665-60, reapproved 1973. Standard Method of Test for Rust-Preventing Characteristics of Steam-Turbine Oil in the Presence of Water.
9. ASTM D 892-74. Standard Method of Test for Foaming Characteristics of Lubricating Oils.
10. ASTM D 943-54, reapproved 1973. Standard Method of Test for Oxidation Characteristics of Inhibited Steam Turbine Oils.
11. ASTM D 974-64, reapproved 1973. Standard Method of Test for Neutralization Number by Color Indicator Titration.
12. ASTM D 1367-64, reapproved 1973. Standard Method of Test for Lubricating Qualities of Graphites.
13. ASTM D 1401-67, reapproved 1972. Standard Method of Test for Emulsion Characteristics of Petroleum Oils and Synthetic Fluids.
14. ASTM D 1741-64, reapproved 1973. Standard Method of Test for Functional Life of Ball Bearing Greases.
15. ASTM D 1796-68, reapproved 1973. Standard Method of Test for Water and Sediment in Crude Oils and Fuel Oils by Centrifuge.
16. ASTM D 1947-73. Standard Method of Test for Load-Carrying Capacity of Fluid Gear Lubricants.
17. ASTM D 2155-66, reapproved 1969. Standard Method of Test for Auto-Ignition Temperature of Liquid Petroleum Products.
18. ASTM D 2266-67, reapproved 1972. Standard Method of Test for Wear Preventive Characteristics of Lubricating Grease (Four Ball Method).
19. ASTM D 2271-66, reapproved 1971. Standard Method of Test for Preliminary Examination of Hydraulic Fluids (Wear Test).
20. ASTM D 2509-68, reapproved 1973. Standard Method of Test for Measurement of Extreme Pressure Properties of Lubricating Greases (Timken Method).
21. ASTM D 2596-69, reapproved 1974. Standard Method of Test for Measurement of Extreme-Pressure Properties of Lubricating Greases (Four Ball Method).

22. ASTM D 2670-67, reapproved 1972. Standard Method of Test for Measuring Wear Properties of Fluid Lubricants (Falex Method).

23. ASTM D 2711-74. Standard Method of Test for Demulsibility Characteristics of Lubricating Oils.

24. ASTM D 2782-74. Standard Method of Test for Measurement of Extreme-Pressure Properties of Lubricating Fluids (Timken Method).

25. ASTM D 2783-71. Standard Method of Test for Measurement of Extreme Pressure of Lubricating Fluids (Four Ball Method).

26. ASTM D 2877-70. Standard Method of Test for Measuring Frictional Properties of Slideway Lubricants. (This method was discontinued in November, 1975.)

27. ASTM D 2882-74. Standard Method for Vane Pump Testing of Petroleum Hydraulic Oils.

9-9 STUDY QUESTIONS AND PROBLEMS

1. Compute the coefficient of friction for a flat metal block with a mass of 250 kg (17.1 slug) which, when pulled on a flat surface, requires a force of 500 N (112.4 lbf) to sustain movement.

2. Compute the pressure and coefficient of friction if the flat block in Problem 1 has an area of 2 m^2 (21.5 ft^2) and is separated by a lubricant film 5.1×10^{-5} m (16.7×10^{-5} ft) thick which has an absolute viscosity of 2×10^{-2} N·s/m^2 (4.2×10^{-4} lb·s/ft^2) and travels at a velocity of 0.5 m/s (1.6 ft/sec).

3. Compute the normal clearance for a 13-cm (5.1-in.) diameter pressure-lubricated shaft that has a bearing length-to-diameter ratio of $L/D = 1.5$.

4. A 10-cm (4-in.) shaft supports a mass of 100 kg (6.8 slug) equally between two pressure-lubricated cylindrical bearings. Assuming a bearing length of $L = 1.5D$, compute the pressure of the lubricant.

5. Compute the tangential force necessary to turn a 6-cm (2.4-in.) shaft 1500 rpm if the cylindrical bearing is pressure-lubricated with an oil that has an absolute viscosity of 2×10^{-2} N·s/m^2 (4.2×10^{-4} lbf·s/ft^2). Assume that $c = D/1000$ and that $L = 1.5D$.

6. If the shaft in Problem 5 supports a load of 50 kg (3.4 slug), compute the coefficient of friction associated with the bearing and oil used.

7. Compute the horsepower loss associated with turning the shaft in Problem 5.

8. If in Eq. (9.12), S_2 has the minimum value of 5×10^{-8} to make the relationship valid, and $D/c = 1000$, compute the minimum coefficient of friction.

9. If a 5-cm (2-in.) cylindrical shaft under a load of 100 kg (6.8 slug) turns at 1500 rpm in a bearing 7.5 cm (3 in.) long that has a clearance of $c = D/1000$, what would be the absolute viscosity of the oil to maintain the f-factor at 0.05?

10. If for a specific oil the relationship $\mu N / P$ should not fall below 5×10^{-6} to maintain a f-factor minimum at 0.04, what will be the bearing clearance-to-diameter ratio?

11. Compute the API degrees at 60°F of a fluid with a Sg of 0.93 at 60°F.

12. Compute the viscosity index of an oil that has an SSU viscosity of 80 at 210°F and a viscosity of 1000 at 100°F.

13. What are the differences among the following: flash point, fire point, and auto ignition temperature?

14. What are the differences among oxidation, rusting, and foaming in fluids?

15. What is the difference between a strainer and a filter?

10

FLUID MACHINERY

10-1 INTRODUCTION

Fluid machinery adds, subtracts, and transmits power, using working fluids such as water, oil, and air. Turbines, pumps, motors, valving, and accessory equipment provide the means to harness the working fluid for productive purposes. Machinery is designed to accomplish specific purposes, given the engineering and financial constraints that are consistent with the available technology. Some branches of the technology, such as hydroelectrics, have a long history worldwide, while others such as aerospace applications, have been pioneered only recently by a few countries.

Fluid machinery and related industries may be classified in a number of ways: by the amount of power produced, consumed, or transmitted; by capital investment and size of the industry; by the type of mechanism used to harness the working fluid such as positive displacement (static) or kinetic (dynamic) equipment; by the fluid used; and by major industry, such as utilities, manufacturing, construction, and transportation. It is also common to subdivide industrial applications into such areas as stationary and mobile equipment, or by product produced, such as metal-forming, earth-moving, and meat-packing equipment. Classifications are usually derived upward in hierarchy from the producer's perspective to fit ultimately into systems that have nationwide or worldwide acceptance, such as the Standard Industrial Classification[1] of products.

---

[1]The Office of Management and Budget, *Standard Industrial Classification Manual* (Washington, D.C.: U.S. Government Printing Office).

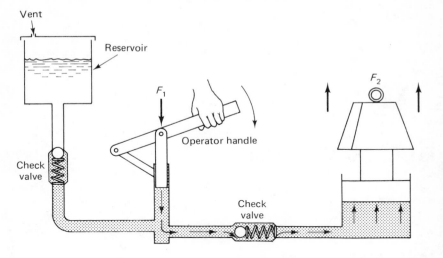

Fig. 10-1   Transmission and multiplication of force

10-2   HOW FLUIDS TRANSMIT POWER

Fluids transmit power by the action of the fluid under pressure. Power in this case is defined as work per unit time. Where the power being transmitted through a system is constant and losses are ignored, multiplication of force is gained at the expense of time required to accomplish the task. Figure 10-1 illustrates the transmission and multiplication of force by a fluid under pressure using a positive displacement pump and cylinder. The small piston delivers fluid from the reservoir to the large cylinder, which in this case is used to lift a load. The multiplication of force equals the ratio of the area of the larger piston output cylinder to the area of the smaller piston input cylinder. However, the distance traveled by the load compared to the distance traveled by the input pump piston is the reciprocal of this relationship. The pressure in the fluid during this time is the same throughout the system. Thus, the conditions of Pascal's law[2], that the pressure acts undiminished in all directions, and the law of conservation of energy are both met. In notation, since the pressure is constant and proportional to the load (force),

$$p = \frac{F}{A}$$

then

$$\frac{F_1}{A_1} = \frac{F_2}{A_2} \tag{10.1}$$

[2]Blaise Pascal (1650).

EXAMPLE 10-1

If the pump piston is half the size of the output cylinder piston, what will be the multiplication of force at the output and how far must the pump piston travel in relation to the output cylinder piston?

SOLUTION

Rearranging Eq. (10.1) to solve for F_2/F_1, we have

$$\frac{A_2}{A_1} = \frac{F_2}{F_1}$$

and since the diameter of the output piston A_2 is twice that of the input piston A_1, the ratio of their areas is 4:1. Thus

$$\frac{F_2}{F_1} = 4:1$$

The distance traveled by the load piston compared to the input piston is the reciprocal of this, or 1:4.

The power transmitted by the system if the load (and thus pressure) is constant is a function of the fluid flow rate. The formula that expresses the relationship between load, flow rate, and power is given in several forms, depending usually upon the application, which varies by industry. Subsequent sections will address specific formulas with related applications.

Figure 10-2 illustrates a positive displacement hydraulic gear pump and system to transmit power to move a cylinder in both directions. The components of the system are the reservoir, the gear pump, the pressure relief valve to prevent overloading, the four-way control valve, the double-acting cylinder, and interconnecting tubing. When the control valve is in the neutral position, the oil circulates through the pump, center spool, and back to the reservoir. Shifting the control valve spool downward directs the fluid to the cap end of the cylinder, thus raising the load, and permits oil at low pressure from the rod end of the cylinder to return to the reservoir. Shifting the spool upward directs fluid to the head end of the cylinder and allows the load to return to its original position as oil from the blank end of the cylinder returns through the control valve to reservoir. The relief valve opens when the cylinder reaches either the extremity of its stroke under power or an immovable obstruction, allowing the fluid delivered from the positive displacement gear pump to return to the reservoir without damaging components or plumbing.

Positive displacement systems are characterized by the transfer of a constant volume of fluid per cycle or flow rate per unit time. When the

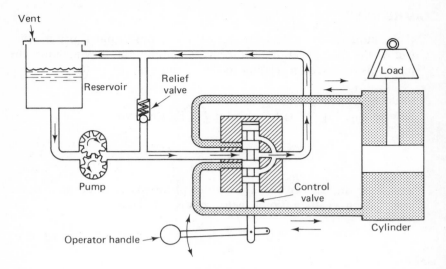

Fig. 10-2 Hydraulic system

internal volume of the pump or motor cannot be adjusted, it is considered to have a fixed displacement. If, on the other hand, the pump or motor internal geometry and volume can be increased and decreased, it is said to have a variable displacement. Systems equipped with variable displacement pumps have the advantage of being able to adjust to varying conditions of load and speed. Although the drive speed of the pump and motor can also be varied to increase or decrease the volume flow rate for positive displacement pumps and motors, this is a less desirable alternative to varying the internal displacement of the pump and/or motor to change performance characteristics.

There are several pump and motor mechanisms of the positive displacement type, including gears, vanes, rotating lobes, pistons, and screws. Each of these in turn has variations in the mechanical configuration and arrangement of the moving members.

Figure 10-3 illustrates four positive displacement fluid mechanisms that represent configurations of pumps or motors. In Fig. 10-3(a), pumping action occurs when the input drive shaft causes one gear to rotate the other. This action, in turn, causes fluid to be displaced from the inlet port to the outlet port as follows: The gear teeth seal as one rotates the other. As the teeth part on the inlet side of the pump, increases in the volume of the inlet chamber cause a slight vacuum to form. The rotating gear teeth thus transfer fluid drawn in and trapped between the gear teeth around the outside periphery of the gears to the high-pressure outlet port. Meshing of the teeth near the outlet chamber reduces the cavity volume by an amount

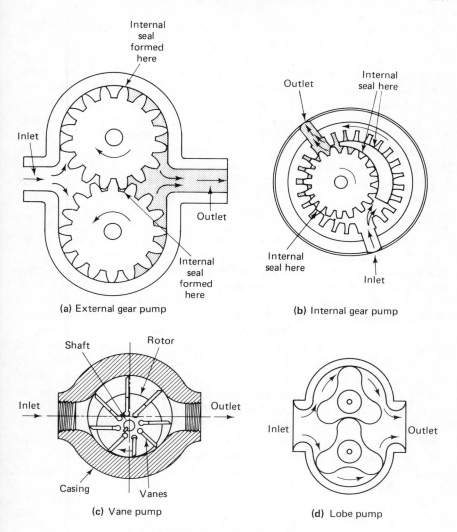

(a) External gear pump

(b) Internal gear pump

(c) Vane pump

(d) Lobe pump

Fig. 10-3 Positive displacement mechanisms

equal to the displacement between the teeth as they come together. This forces fluid from the outlet cavity and port at system pressure, which is dependent upon the load resistance. Another arrangement for a gear pump and motor is illustrated in Fig. 10-3(b). Internal gear pumps and motors operate similarly to external gear pumps, except the internal spur gear drives the outside ring gear, which is set off center. Between the two gears on one side is a crescent-shaped spacer around which oil is carried. The inlet and outlet ports are located in the end plates between the place where the teeth mesh and the ends of the crescent-shaped spacer. In operation, fluid is directed from the inlet to the outlet port. As a pump, rotation of the internal gear causes the teeth to unmesh near the inlet port, increasing the cavity volume and forming a slight vacuum. The fluid is trapped between the internal and external gear teeth on both sides of the crescent-shaped spacer and carried from the inlet to the outlet cavity of the pump. Meshing of the gear teeth reduces the volume in the high-pressure cavity near the outlet port, and fluid is forced from the outlet port at system pressure. Figure 10-3(c) illustrates a vane pump or motor. Essential components include the inlet and outlet ports, rotor, sliding vanes, and stationary cam ring. The rotor, vanes and cam ring are replaceable as a unit. When operating as a pump, the radially sliding vanes in the rotor alternately increase and decrease the crescent-shaped space between the rotor and the cam ring. The vanes in the rotor slots alternately extend and retract as the rotor turns through each half revolution, trapping fluid from the inlet port and transferring it to the outlet port. An increase in pump cavity volume and suction at the inlet port causes fluid to enter the inlet. At the crossover point, the fluid cavity is at a maximum, and continued rotation causes fluid to be expelled at the outlet port at system pressure. The lobe pump shown in Fig. 10-3(d) transfers fluid in much the same way as a gear pump, except that the close-fitting lobes are synchronized to maintain clearance between the lobes. Rotation transfers the fluid around their outside peripheries. Because of their high volume and speed characteristic, they have wide application in pumping air as supercharges and transferring products that may be suspended in air streams.

Two characteristics that describe the performance of positive displacement rotary pumps and motors are flow rate Q and torque T. The theoretical flow rate Q_t through a positive displacement pump or motor can be computed from the displacement per revolution V and speed of rotation N. That is,

$$Q_t = VN \tag{10.2}$$

The equation for theoretical torque T_t is

$$T_t = \frac{Vp_d}{2\pi} \tag{10.3}$$

where p_d is the positive pressure difference between the inlet and outlet ports. It is assumed, of course, that in Eqs. (10.2) and (10.3) there is no leakage, slippage, vaporization of the fluid, or friction. This is not the case, of course, and each of these constitutes a portion of the total losses. Whereas slippage and leakage can be measured in the laboratory directly with relative ease, other losses are usually computed from the overall efficiency, which is the quotient of the output to input, or the product of each of the separate efficiencies. That is, where e_o is the overall efficiency,

$$e_o = \frac{\text{output}}{\text{input}} \times 100 \qquad (10.4)$$

or

$$e_o = (e_1)(e_2)(e_3)\ldots(e_n) \qquad (10.5)$$

where (e_1), (e_2), etc. are volumetric, mechanical, and other contributing efficiencies.

EXAMPLE 10-2

A pump with a positive displacement of 38 cm³/rev (2.32 in³/rev) delivers 65 l/min (17.2 gpm) of hydraulic fluid at an increase in pressure of 103 bars (1493.5 psi). If the shaft speed is 1800 rpm, compute the volumetric efficiency and the torque applied to the shaft.

SOLUTION
The relationship between displacement V_d, flow rate Q, and shaft speed N is given from

$$V_d = \frac{Q}{N}$$

From the statement of the problem, theoretical delivery Q_t is

$$Q_t = V_d N = (38\ \text{cc/rev})(1/1000\ \text{l/cc})(1800\ \text{rev/min}) = 68.4\ \text{l/min}$$

Thus volumetric efficiency is computed from

$$e_V = \frac{Q}{Q_t} = \frac{(65\ \text{l/min})}{(68.4\ \text{l/min})}(100) = 95\ \text{percent}$$

and from Eq. (10.2), actual torque T_a applied would be

$$T_a = \frac{V p_d}{2\pi} = \frac{(38 \times 10^{-6}\ \text{m}^3)(103 \times 10^5\ \text{N/m}^2)}{(2)(3.14)} = 62\ \text{N} \cdot \text{m}\ (46\ \text{lbf} - \text{ft})$$

Volumetric efficiency, usually computed as a percentage of the theoretical volume flow, is also computed as the actual amount that leaks past the close-fitting parts of the pump or motor. This is termed slip s and is computed from

$$s = Q_t - Q \qquad (10.6)$$

and as a percentage of the theoretical volume flow Q_t

$$s = \frac{(Q_t - Q)}{Q_t} \times 100 \qquad (10.7)$$

or

$$s = \left(1 - \frac{Q}{Q_t}\right) \times 100 \qquad (10.8)$$

Nonpositive displacement pumps and motors are divided into three main categories, depending upon the configuration of the runner element and how it interacts with the fluid. These are centrifugal or radial flow, propeller or axial flow, and mixed flow. The Pelton wheel, which is an example of a axial flow turbine, is described in Chapter 8. Figure 10-4(a)

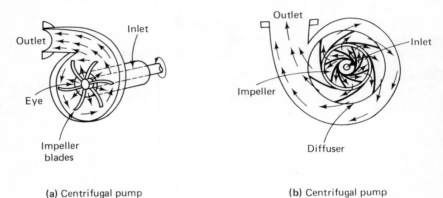

(a) Centrifugal pump   (b) Centrifugal pump

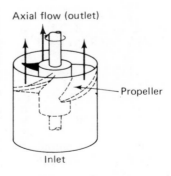

(c) Propeller

Fig. 10-4   Nonpositive displacement mechanisms

illustrates a simple centrifugal pump used to transfer large volumes of fluids at relatively low pressures. Notice that there is only one moving part and that this is a relatively simple impeller mechanism. Advantages of these pumps include low production costs, simplicity of operation, high reliability, low maintenance, and the ability to transfer nearly all fluids without damage to internal parts. Because the inlet and outlet passages are connected, however, centrifugal pumps are not self-priming except with air, and must be positioned below the level of the liquid or primed to initiate pumping action. While most have large internal clearances, some water circulating units use flexible impeller blades to reduce internal clearances and increase pumping efficiency. Because these units transfer high volumes at low pressures, they have wide application as supercharging pumps, liquid transfer pumps in low-pressure hydraulic applications requiring high fluid flow rates, such as traverse feed mechanisms and sump pumps. The parts of a centrifugal pump include an inlet port, a pumping chamber shaped as an involute, the drive shaft and impeller, and the outlet port. Notice in Fig. 10-4(a) that the blades are curved opposite to the direction of rotation to increase efficiency. Stationary defuser blades are sometimes attached to the pump housing to redirect the fluid in such a way as to reduce the velocity and internal clearances. This has the effect of increasing the capacity of the pump to develop pressure [Fig. 10-4(b)]. Centrifugal pressure builds toward the outside and is increased by velocity reduction in the expanding section of the snail-shaped duct leading to the outlet.

Figure 10-5 illustrates a vector diagram for a centrifugal pump where fluid enters near the hub at a velocity $v_1 = 0$ and exits at the periphery from

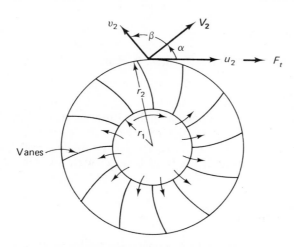

Fig. 10-5 Vector diagram for centrifugal pump

the trailing edge of the vanes set at an angle β (Greek letter beta) at a velocity v_2. From the momentum equation, the force exerted on the fluid is

$$F = \rho Q V_2 \tag{10.9}$$

where ρ is the density of the fluid, Q is the flow rate, and V_2 is the resultant velocity vector of the fluid as it leaves the impeller. When F is positive, the impeller is expending work on the fluid. When F is negative, the machine is a turbine and the work is done by the fluid on the rotating member.

The component of the force F_t that acts at a tangent to the rotor in the direction of rotation is

$$F_t = \rho Q V_2 \cos \alpha \tag{10.10}$$

from which the torque T is

$$T = F_t r_2 = \rho Q V_2 (\cos \alpha) r_2 \tag{10.11}$$

and the power expended is

$$P = T \omega = \left[(\rho Q V_2 (\cos \alpha) r_2) \right] = (\rho Q V_2 \cos \alpha)(r_2 \omega) \tag{10.12}$$

But the angular velocity equals

$$\omega = \frac{u_2}{r_2}$$

and

$$u_2 = \omega r_2 \tag{10.13}$$

So that by substitution

$$P = \rho Q u_2 V_2 \cos \alpha \tag{10.14}$$

EXAMPLE 10-3

A 15-cm (5.91-in.) diameter impeller on a centrifugal pump rotates at 1750 rpm (Fig. 10-6). If the relative fluid velocity at a blade angle $\beta = 120$ deg is 15 m/s (49.2 ft/s) and the angle $\alpha = 45°$, compute the pressure at the outlet if the fluid is water with $\rho = 1000$ N·s²/m⁴ (1.94 slugs/ft³).

SOLUTION
From Eq. (10.13)

$$u_2 = \omega r_2 = \left(\frac{2\pi N}{60} \right)(15 \times 10^{-2} \text{ m}) = 27.47 \text{ m/s}$$

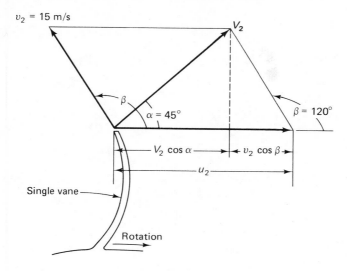

Fig. 10-6   Example 10-3

From the vector diagram in Fig. 10-6, it can be seen that

$$V_2 \cos \alpha = u_2 - v_2 \cos \beta = (27.47 \text{ m/s}) - \left[(15 \text{ m/s})(0.5) \right] = 19.97 \text{ m/s}$$

Since from Eq. (5.6)

$$P = pQ$$

Eq. (10.14) can be written in the form

$$P = pQ = \rho Q u_2 V_2 \cos \alpha$$

and

$$p = \rho u_2 V_2 \cos \alpha \tag{10.15}$$

Solving for the pressure, we obtain

$$p = (1000 \text{ N} \cdot \text{s}^2/\text{m}^4)(27.47 \text{ m/s})(19.97 \text{ m/s}) = 549 \text{ kPa} = (79.5 \text{ psi})$$

Figure 10-7 illustrates a vector diagram for a radial flow turbine runner. Fluid enters the vanes around the outside periphery at an angle (β_1) and relative velocity v_1 and exits near the hub from the trailing edge of the vanes at an angle (β_2) and relative velocity v_2. From the momentum equation and Fig. 10-7 which shows the absolute velocities V_1 and V_2

$$F = \rho Q (V_2 - V_1)$$

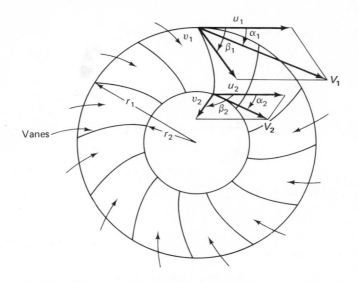

Fig. 10-7 Vector diagram for radial flow turbine runner

or

$$F = -\rho Q(V_1 - V_2) \tag{10.16}$$

Notice that because work is being done on the turbine runner the sign is negative, and that since V_2 has the same direction as V_1, it indicates that unavailable energy remains in the fluid after passage through the vanes, thus reducing V_1 and the force that drives the runner.

The component of the force F_t that acts at a tangent to the runner in the direction of rotation is

$$-F_t = \rho Q(V_1 \cos\alpha_1 - V_2 \cos\alpha_2) \tag{10.17}$$

from which the torque T is

$$T = -F_t(r_1 - r_2) = \rho Q(r_1 V_1 \cos\alpha_1 - r_2 V_2 \cos\alpha_2) \tag{10.18}$$

and by substitutions made in Eqs. (10.12) and (10.13), the power is computed from

$$P = \rho Q(u_1 V_1 \cos\alpha_1 - u_2 V_2 \cos\alpha_2) \tag{10.19}$$

where point 1 is upstream and point 2 is downstream.

EXAMPLE 10-4

A turbine runner (Fig. 10-8) rotates at 220 rpm and discharges 3.5 m³/s (123.6 lbf³/s). Water with $(\rho = 1000 \ \text{N} \cdot \text{s}^2/\text{m}^4)(1.94 \ \text{slugs/ft}^3)$ enters from the outside periphery of the turbine at 25 m/s and exits

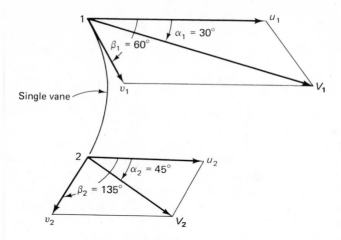

Fig. 10-8   Example 10-4

through the center at 10 m/s so that $\alpha_1 = 30$ deg, $\beta_1 = 60$ deg, $\alpha_2 = 45$ deg, $\beta_2 = 135$ deg, $r_1 = 0.75$ m (2.46 ft), $r_2 = 0.60$ m (1.97 ft). Neglecting losses, compute the power developed by the turbine wheel.

SOLUTION
From the vector diagram in Fig. 10-8, several preliminary calculations must be made before computing the power from Eq. (10.19). From Eq. (10.13) and beginning at point 1

$$u_1 = \omega r_1 = \left(\frac{2\pi N}{60}\right)(0.75 \text{ m}) = 17.27 \text{ m/s}$$

and

$$u_2 = \omega r_2 = \left(\frac{2\pi N}{60}\right)(0.60 \text{ m}) = 13.82 \text{ m/s}$$

From the vector diagram in Fig. 10-8 it can be seen that

$$V_1 \cos\alpha_1 = u_1 + v_1 \cos\beta_1 = (17.27 \text{ m/s}) + (25 \text{ m/s})(0.5) = 29.77 \text{ m/s}$$

and

$$V_2 \cos\alpha_2 = u_2 - v_2 \cos\beta_2 = (13.82 \text{ m/s}) - (10 \text{ m/s})(0.707) = 6.75 \text{ m/s}$$

Finally, solving for the power using Eq. (10.19), we have

$$P = \rho Q (u_1 V_1 \cos\alpha_1 - u_2 V_2 \cos\alpha_2)$$

and

$$P = (1000 \text{ N} \cdot \text{s}^2/\text{m}^4)(3.5 \text{ m}^3/\text{s})[(17.27 \text{ m/s})(29.77 \text{ m/s})$$

$$- (13.82 \text{ m/s})(6.75 \text{ m/s})] = 1\ 472\ 975 \text{ J/s} = 1974 \text{ hp}$$

The configuration of the turbine runner determines its type and operations. Figure 10-9 illustrates the components of a turbine which surround a radial flow Francis runner. Water is directed to the runner through the snail-shaped scroll case, which imparts a whirl action to the flowing fluid. The turbine runs completely full from the entrance to the exit of the draft tube. Adjustable guide vanes called *wicket gates* located around the periphery of the runner direct the fluid in a tangential direction, which causes the fluid to enter and pass across the axial blades smoothly without turbulence or separation. The angle of the fluid can be changed to best suit the speed, load, and available head. After passing through the runner, the fluid is directed down the tapered draft tube, which both decreases the velocity and increases the pressure drop across

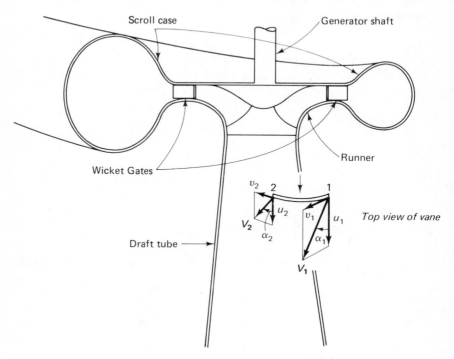

Fig. 10-9 Radial flow (Francis) turbine

Fig. 10-10 *700 MW turbines at Grand Coulee Dam (Courtesy of Allis-Chalmers)*

the runner, thus raising its efficiency. The vector diagram is a top view of how fluid is directed across the vanes.

Figure 10-10 illustrates installation of the Francis runner for the 700 MW turbines No. 22, 23, and 24 at Grand Coulee Dam (Washington state). The 408-metric-ton (450-ton) runner, which has an outside diameter of 9.9 m (32.5 ft), rotates at 85.7 rpm under a head of 87 m (285 ft).

The Kaplan axial flow turbine illustrated in Fig. 10-11 directs the fluid through the scroll case to the wicket gates, which then impart a tangential direction and swirl action to the fluid above the propeller blades. Propeller runners have four to six blades. The downward swirl or vortex action above the runner causes the fluid to enter the blades' smoothly, moving the runner in the direction shown by the side view vector diagram. The draft tube increases the pressure drop across the runner by reducing the velocity

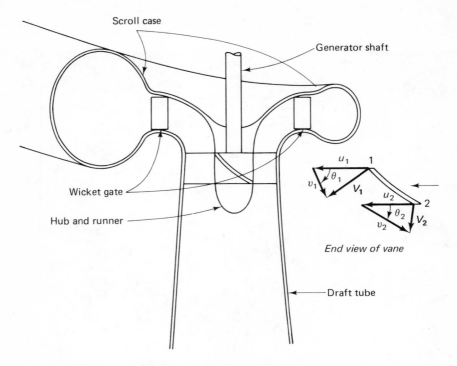

Fig. 10-11   Axial flow (propeller) turbine

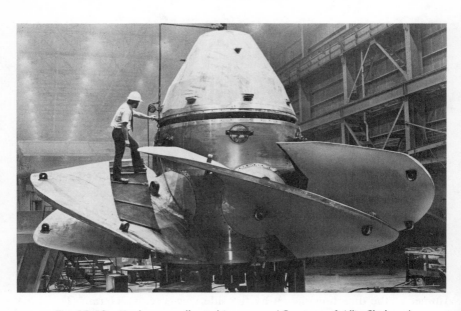

Fig. 10-12   Kaplan propeller turbine runner (*Courtesy of Allis-Chalmers*)

head and allows the fluid to dissipate the remaining swirl of the fluid from the runner.

Figure 10-12 illustrates a Kaplan propeller undergoing testing during final assembly.

10-3 SPECIFIC SPEED N_s AND SPEED FACTOR ϕ

Specific speed is used to describe the performance characteristics of centrifugal pump impellers and turbine runners. Traditionally, it is given for centrifugal pumps as

$$N_s = \frac{N\sqrt{Q}}{h^{3/4}} \text{ (pumps)} \tag{10.20}$$

where flow rate Q in ft^3/s at a given head h in ft are primary considerations, and N is the speed in rpm at maximum efficiency for a given head.[3]

For turbine runners, since the primary interest is in potential power HP from an available head h in ft, and since power is proportional to hQ

$$N_s = \frac{N\sqrt{HP}}{h^{5/4}} \text{ (turbines)} \tag{10.21}$$

In terms, then, the specific speed for a centrifugal pump is the speed at which it would deliver a volume flow rate of $Q = 1$ ft^3/s against a head of $h = 1$ ft. For turbines, it is the speed at which an available head of ($h = 1$ ft) would generate one horsepower (HP). In both cases, N_s is in English units.

Because N_s remains nearly constant for a specific type of impeller or turbine runner regardless of size, it is useful in making comparisons between different types of units of comparable size, or runners of the same type but of different sizes. This is common practice in hydraulic laboratories, where scale models are installed in a test system to predict the operating characteristics of a proposed full-scale prototype. In notation

$$\frac{N\sqrt{HP}}{h^{5/4}} = \frac{N'\sqrt{HP'}}{(h')^{5/4}}$$

where the speed, horsepower, and head of the scale model have been given prime notation. Thus, the horsepower for a full-scale prototype may be accurately predicted from a scale model by using

$$HP = \left(\frac{N'}{N}\right)^2 \left(\frac{h}{h'}\right)^{5/2}(HP')$$

---

[3]If one distinguishes between pumps and fans, Q is in gallons per minute for pumps and in cubic feet per minute for fans. Some authorities use unit discharge for pumps or fans as a unit time of one minute.

Other similar relationships can also be developed for both pumps and turbines.

The speed factor ϕ is defined as the ratio of the peripheral velocity of the pump impeller or turbine runner u, to the velocity of the fluid leaving or entering the vanes ($v = \sqrt{2gh}$). In notation

$$\phi = \frac{u}{\sqrt{2gh}} = \frac{\pi DN}{60\sqrt{2gh}} = \frac{\omega r}{\sqrt{2gh}} \qquad (10.22)$$

where h is the head generated by the pump impeller or available head to the turbine runner, D or r are measured to the outside of the runner, and N is the speed in revolutions per minute. Speed factor depends upon specific speed at most efficient operating conditions.

10-4   FLUID POWER SYSTEMS

A fluid power system is one that transmits and controls power through the use of a pressurized fluid, i.e., gas or liquid, within a closed circuit.

By product, fluid power systems include pumps, motors, cylinders, valves, and related components. Intensifiers, accumulators, reservoirs, logic control units, filter assemblies, tubing and fittings, and gauges comprise most of these.

Fluid power systems provide mechanical power to move parts of stationary industrial equipment, such as machine tools, into precise positions. Fluid power is applied to aerospace technology to control the landing gears and flaps of aircraft. Missiles and rockets are elevated into firing position by the use of fluid power systems. Fork lift trucks and other materials handling equipment employ pressurized fluids, as does large construction, farm, and mining machinery.

The pressures developed by fluid power systems, to 1 MPa (10 bars)(145 psi) for most air systems and between 10 and 34.5 MPa (100 and 345 bars)(1500–5000 psi) for hydraulic systems, require that these products be manufactured under high standards of precision and quality control. Machine tools and much of the other equipment that these components drive must be equally accurate and precise.

A recent survey of the industry in the United States indicates that pumps, motors, and cylinders account for approximately 60 percent of total production in a world market that is estimated to be 3.4 billion dollars, divided evenly between the United States and the rest of the world. Western Europe and Japan are estimated to account for about 90 percent of that portion of the international market in hydraulics with 65 percent and 25 percent, respectively.[4,5]

By end use, fluid power equipment is classified as stationary, mobile, aerospace, marine, and special military. Stationary or industrial equipment, as it is often called, is used in fabricating, manufacturing, and processing operations. It includes machine tools, forging machinery, conveyors, and related equipment. Aerospace equipment is used in missiles and spacecraft, and on aircraft, including military, commercial, and private. Marine applications include ships, boats, and other watercraft. Special military equipment is used by the armed services in such applications as tanks, launchers, and radar and ordnance equipment.

The manufacturing and machine tool industry is very dependent on hydraulics to provide the power and close tolerance necessary for controlled production. Figure 10-13 illustrates a 2000-ton forging press, one of the largest self-contained presses in existence. Hydraulic fluid is supplied by 20 pumps, each of which has a capacity of 132 l/min (35 gpm). At an operating pressure of 13.8 MPa (2000 psi), the press consumes 800 horsepower. The press is driven by fluid supplied from a central hydraulic pumping system consisting of ten double-end electric motors driving the 20 hydraulic pumps.

Air compressors are classified as positive or nonpositive displacement, single or multiple stage, reciprocating or rotary, by pressure level and by delivery or rated horsepower. A modern compressor room housing two large stationary Y-type two-stage water-cooled industrial compressors is shown in Fig. 10-14. While positive displacement compressors are designed with sliding or rotary vanes as well as lobe impeller types, most large stationary industrial compressors are of the reciprocating piston type, which arranges the pistons in single or double staging to compress the air. There is a practical limit to the pressure that can be developed in single-stage piston-type compressors. This is reached at about 1 MPa (10 bars)(145 psi), where the compression ratio, compressing chamber size, and heat of compression act to impede efficient pumping action. Staging has the effect of increasing pumping efficiency by dividing total pressure between two or more stages and reducing the heat associated with inefficiency by intercooling between stages. Multistage piston compressors pipe the exhaust of each succeeding cylinder to the next, which has a

[4]U.S. Department of Commerce, *Fluid Power Systems Equipment, Production and End Use* (A survey Report). Superintendent of Documents, U.S. Government Printing Office, Washington, D.C. 20402, 1975.
[5]Illinois Institute of Technology, *Proceedings of the Thirty-First National Conference on Fluid Power, Worldwide Hydraulic Pump/Motor Needs* (Similarities and Differences with U.S. Needs), 1975, pp. 630–640.

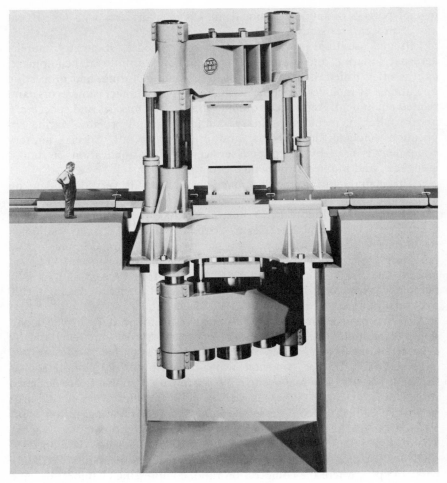

Fig. 10–13   2000-ton self-contained forging press (*Courtesy of Abex Corporation, Denison Division*)

smaller size, thereby stepping the pressure. Two-stage compressors have a pressure range between 0.7–1.7 MPa (100–250 psi).

Transportation systems provide examples of the most varied uses of fluid power. Road and off-the-road vehicles are typically suspended by pneumatic tires. Air-assisted brakes, both high-pressure and vacuum assisted, are standard equipment. Power steering systems, either of the hydraulic assist or full power steering type, are also standard and relieve the operator of the fatigue commonly associated with physical exertion over long periods of time. Hydrostatic transmissions are now common on

Fig. 10-14   Stationary Y-type air compressor (*Courtesy of Worthington*)

all manner of vehicles. Suspension systems are dampened with hydraulic shock absorbers, and some combine pneumatics by using the compressible nature of gases as the basis for air-oil suspension systems. Hydraulic wheel drive motors are recent to the transportation line of vehicles. They give almost unlimited flexibility to the designer by permitting the power plant to be located centrally to power wheels through high-pressure lines, where they are required to support the load and provide traction in propelling the vehicle with a variety of stepless speeds. Figure 10-15 illustrates a large off-the-road truck that incorporates many of these features. With a railroad diesel engine rated at 2475 hp, this vehicle is capable of transporting loads of 124–161 m^3 (162–211 yd^3) for payloads up to 213 metric tons (235 tons). Tandem axle air-hydraulic suspensions are hydraulic coupled to perform as a walking beam for equalized tire loads and shock loads. Land mobile equipment is now estimated to comprise approximately 50 percent of the dollar volume end use of fluid power products.

By components, industry production in the United States is greatest in four areas: valves (33 percent), cylinders (29 percent), pumps (20 percent), and motors (11 percent). Hydraulic and pneumatic valves control the pressure, flow rate, and direction of fluid in the system. They are operated both manually and remotely by fluid or electrically controlled pilots. Pressure control valves protect the system and work products from damage resulting from peak pressures. They are necessary in systems that use fixed

Fig. 10-15 Off-the-road truck (*Courtesy of WABCO Construction and Mining Equipment Group, An American Standard Company*)

displacement pumps that are not themselves compensated for pressure. Flow control valves control the fluid output flow rate from the pump to the actuators in the system such as cylinders, and consequently control the rate at which work is accomplished. Flow control valves also divide the flow of fluid on a priority or proportional basis to maximize the use of available flow from the system pump to one, two, or more of the several circuits that are working. Directional control valves manage the direction of the fluid, including stopping and starting. In mobile applications, control valves are typically actuated by the physical force provided by an operator. Industrial and other applications make extensive use of solenoid valves and fluid pilot-operated valves, where the pilot is controlled by a solenoid. The internal mechanisms that physically accomplish the valving action use a variety of elements including poppets, diaphragms, flat slides, balls, round shear action plates, and rotating or sliding spools.

Fluid power cylinders exert a linear force and hold it at any specified position indefinitely. Cylinders may be of the single-acting type, which requires the load resistance to return it to the retracted position, as does a car hoist, or of the double-acting type, applying pressurized fluid alternately to both the blank end and rod end. Cylinders are constructed of a cylinder barrel, piston and rod, end caps, and seals. The piston provides the effective area against which fluid pressure is applied and supports the piston end of the rod. The cylinder bore, end caps, ports, and seals maintain a fluid-tight system into which fluid energy is piped. Routing of the fluid determines the direction initiated by the cylinder. Another feature sometimes incorporated in cylinders is a cushion, snubber, or decelerator to reduce shock loads to the mechanism that would be caused by bottoming of the piston in either extreme position. Cylinders in which the rod approaches the size of the bore are called rams, and give maximum support to the load end of the rod. They are used extensively for dump cylinders, automotive hoists, and hydraulic presses.

Fluid power pumps convert mechanical energy into a flow of pressurized fluid in the system. Conversely, fluid power motors convert the flow of pressurized fluid into continuous rotary motion. Pumps and motors may be of the gear, vane, or piston type; the piston category may be divided into axial units, where the pistons reciprocate in the same axis as the shaft, or radial units, where the pistons are arranged at right angles to the shaft. Figure 10-16 illustrates several typical hydraulic units. Gear pumps and motors are usually of the fixed displacement type. Vane and piston pumps have the inherent capability to be of either the fixed or variable displacement type. Vane pumps, for example, can vary the displacement by moving the cam ring from an extreme position off center, at which place the displacement will be maximum, to a position concentric with the rotor, reducing the displacement to zero [Fig. 10-3(c)]. Piston pumps accomplish the same function by varying the piston stroke. If displacement varies with output pressure, which is a function of the load resistance against which the unit is pumping, it is called a pressure-compensated variable displacement unit.

Although the ratio of pumps to motors varies considerably from country to country, the international market places the value estimate at about 3.5 to 1 worldwide. This compares with a U.S. market of 2.8 to 1. Within the pump category by value estimates, gear pumps account for about 25 percent, vane pumps for 25 percent, and piston pumps for 50 percent. By quantity, however, gear units would dominate the market because of their lower unit cost.

Operating temperatures of pumps and motors are between 50°–60°C (122°–140°F) for stationary applications, and 80°–105°C (176°–221°F) for mobile applications. Power steering applications commonly have operating temperatures up to 120°C (248°F). Because high temperatures shorten oil life, the trend worldwide is for lower temperatures, probably because of higher oil prices and maintenance costs. Noise levels are on a downward trend from what is now accepted as 85 dB(A) during an eight-hour period, to the 70–75 dB(A) range on a distance of 7 m. System pressures are

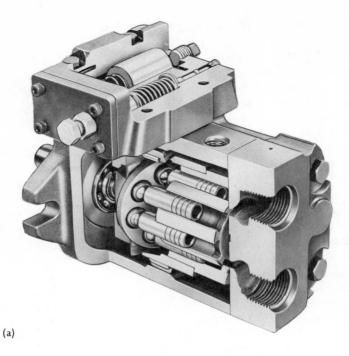

(a)

(b)

Fig. 10-16 Positive displacement hydraulic pumps and motors: (a) Pressure compensated variable delivery axial piston pump (*Courtesy of Hydura Products—The Oilgear Company*); (b) External gear pump (*Courtesy of Webster Electric Company, Inc.*)

(c)

(d)

Fig. 10-16 Positive displacement hydraulic pumps and motors (continued): (c) Variable displacement bent axis piston pump (*Courtesy of Abex Corporation, Denison Division*); (d) Radial piston motor (*Courtesy of Poclain Hydraulics*)

generally expected to increase in the following industries, where pressures are now: machine tool, 5.0–7.0 MPa (725–1000 psi); primary metals, 15.0–25.0 MPa (2175–3625 psi); construction and earthmoving, 17.5–22.5 MPa (2537–3262 psi); and truck and bus, 10.0–14.0 MPa (1450–2030 psi).

10-5   HYDROELECTRIC TURBINES

Hydroelectric turbine installations provide a valuable source of nonpolluting power, and except for high installation costs, the renewable water supply has been considered to be free. Estimates place the total installed power worldwide in excess of 260000 MW[6] with the majority of installations and most of the power being generated by Francis turbines.

The type of unit selected for a particular installation is determined by the available head, flow rate from the source, and quality of the water. While impulse wheels are used for high head installations up to 800 m (2625 ft) and 112 MW (150000 hp), small units have been designed for heads as low as 61 m (200 ft). Francis turbines, sometimes designated as all those of the inward flow type, are used for heads up to 305 m (1000 ft) and 746 MW (1000000 hp). Propeller turbines, both of the fixed pitch and Kaplan type, are regularly chosen for low heads up to 45 m (150 ft) and 168 MW (225000 hp).

Francis-type reversible pump/turbines developed during the 1940's are now receiving widespread attention for pumped storage projects in the Western Hemisphere as well as in other parts of the world. The pump/turbine unit provides stored hydro power for pumping water from a reservoir or supply at one elevation to a storage reservoir at a higher elevation. The turbine can thus use this potential energy by developing power as the flow passes back from the storage reservoir to the lower pond. Rotation is in one direction when it is operating as a pump and in the opposite direction when it is operating as a turbine. The pump/turbine is direct-connected to an electrical machine, which serves as a motor when the unit operates as a pump and as a generator when it operates as a turbine. Basic to the concept is that water is pumped to the upper reservoir when low-cost power is available from the line, and power is generated during peak load periods when it has a higher value. While pumped storage efficiency depends on several variables, it is possible to obtain 2 to 3 kilowatt hours generation for every 3 to 4 kilowatt hours of pumping power. Designs are in use with heads to 366 m (1200 ft).

Thermal generating stations operate most economically when constantly loaded at the most efficient output. Thus pumped storage can be

[6]Van Der Leeden, Frits. *Water Resources of the World: Selected Statistics* (Port Washington, N.Y.: Water Information Center, Inc., 1975), p. 470.

HYDRAULIC TURBINE GENERATOR UNIT

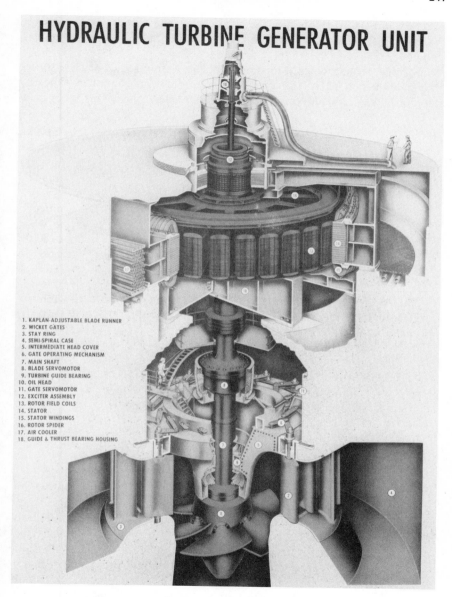

1. KAPLAN-ADJUSTABLE BLADE RUNNER
2. WICKET GATES
3. STAY RING
4. SEMI-SPIRAL CASE
5. INTERMEDIATE HEAD COVER
6. GATE OPERATING MECHANISM
7. MAIN SHAFT
8. BLADE SERVOMOTOR
9. TURBINE GUIDE BEARING
10. OIL HEAD
11. GATE SERVOMOTOR
12. EXCITER ASSEMBLY
13. ROTOR FIELD COILS
14. STATOR
15. STATOR WINDINGS
16. ROTOR SPIDER
17. AIR COOLER
18. GUIDE & THRUST BEARING HOUSING

Fig. 10-17 Hydraulic turbine generator unit (*Courtesy of Allis-Chalmers*)

used to handle load fluctuations in the daily, weekly, or seasonal cycles by absorbing off-peak loads and supplying additional energy during peak loads.

The speed N at which a turbine rotates is a function of its size, power, frequency of the current f in cycles per second, and the number of poles p that can be economically mounted on the spinning rotor. This is given by

$$N = \frac{120 f}{p} \qquad (10.23)$$

where the frequency f is 50 cps in Europe and 60 cps in North America. There is an economy to be realized by having the runner turn as fast as possible to reduce the size of turbine components and the generator itself. Figure 10-17 illustrates a complete turbine generator unit, showing the components of the Kaplan runner, wicket gates and scroll case, and spinning rotor.

EXAMPLE 10-5

The 600-MW turbine units at Grand Coulee Dam rotate at 100 rpm. How many poles must the rotor of the alternator have in order to generate 60 cps current?

SOLUTION
From Eq. (10.23)

$$p = \frac{120 f}{N} = \frac{(120)(60 \text{ cps})}{(100 \text{ rpm})} = 72 \text{ poles}$$

Notice that to be dimensionally correct the constant (120) actually represents (2×60) or 2 poles/cycle times 60 seconds/minute.

The efficiencies of hydroelectric plants, such as the representative unit shown in Fig. 10-18, depend upon several contributing factors. These include how well the water traverses the unit, the mechanical characteristics of the runner mechanism, and losses between the mechanical shaft rotation and conversion to electricity in the generator unit. With these associated losses, hydroelectric generation plants still provide the highest available efficiencies from all practical means of power generation, usually on the order of 90 percent. Although the number of units produced and installed has decreased over time, individual unit capacity and size have increased dramatically since 1960, from about 150 MW(200000 hp) then for large units to 746 MW(1000000 hp) now for the latest three units installed at Grand Coulee Dam. Pump/turbines operating as pumps function at about 65 percent efficiency.

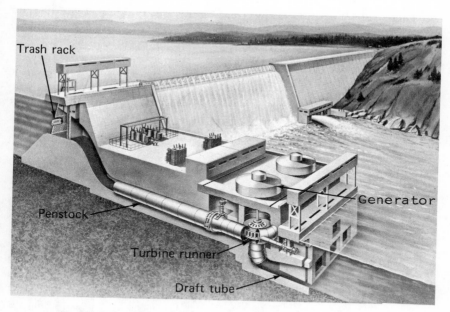

Fig. 10-18 Hydroelectric installation (*Courtesy of Allis-Chalmers*)

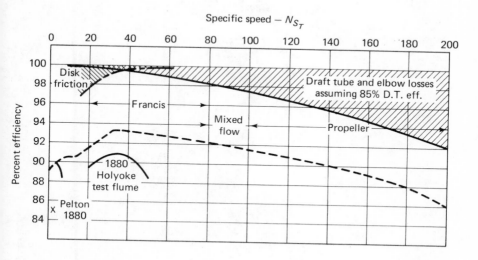

Fig. 10-19 Efficiencies for various types of turbines (*Courtesy of Allis-Chalmers*)

There are a number of ways to compute and interpret losses in fluid machinery, including plotting efficiency e (ordinate) against (abscissas) speed N, capacity Q, power P or percentage of rated power at best efficiency, and specific speed N_s. Figure 10-19 gives some historical perspective to the plot of percentage of efficiency against specific speed for the various types of turbines. Since the 1880's, turbine efficiencies have increased over an expanded range of higher specific speeds. Notice that as specific speed increases, both the flow rate and velocity of the discharge increase, giving more importance to the function of the draft tube, which

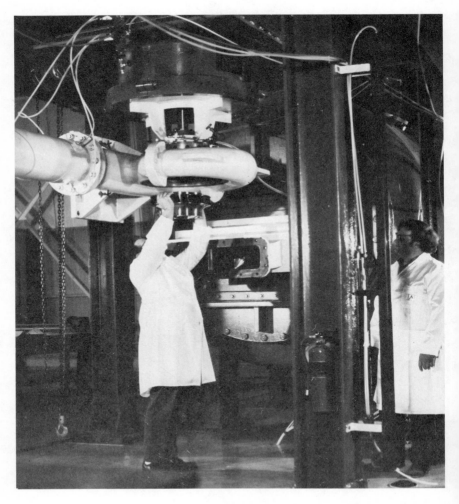

Fig. 10-20   Model turbine test setup (*Courtesy of Commercial Shearing, Inc.*)

must recover the velocity head for an efficient unit. While mechanical friction and runner losses remain almost constant for a specific type of turbine runner, hydraulic losses in the wicket gates, runner blades, and draft tube increase with specific speed, flow rate, and part-load operation.

Estimated losses associated with the entire power plant occur at the trash racks, intake, and penstock (1.0 percent), through the turbine (4–10 percent), and through the generator itself (3 percent at full load, 30–10 percent at $\frac{1}{4}$ to $\frac{3}{4}$ load). Thus the product of these efficiencies yields an overall plant efficiency somewhat lower than that exclusively associated with the turbine runner.

The information used to design installations is derived from models that are tested in laboratories, from computer studies, and from experience gained from installations worldwide by several manufacturers. Remarkable accuracy has been obtained by using the model method to design prototypes with predicted performance characteristics usually within 1–2 percent. This is contrasted with methods and difficulties encountered in monitoring after installation full-size prototypes, which are more expensive to produce and which rarely yield more accurate results. There is also the obvious advantage of defining performance before the prototype is built when the option to change or modify the design is still available. Figure 10-20 illustrates the installation of a model turbine in a test setup in a modern hydroelectric turbine laboratory.

10-6 TESTING AND REPORTING PROCEDURES

The fluids technician[7] is frequently required to assemble an apparatus or equipment, to conduct a testing procedure, and then to report pertinent findings to superiors or clients. Sometimes, conclusions and recommendations are made that affect the acceptance or rejection of a product, material, or procedure. Many components, for example, are tested to see if they meet both the specifications and service life as advertised by a supplier, and this affects whether or not they are used by a manufacturer as part of a larger system. New components also undergo exhaustive testing by developing companies. Figure 10-21 illustrates a technician making routine tests on a hydraulic valve, using specially designed testing equipment. Frequently, such specialized tests and equipment generate standards for the industry, particularly where there is proprietary interest in the technology and state of the art on the part of companies that have developed these components. After use and acceptance by the industry,

---

[7]The U.S. Department of Labor, *Dictionary of Occupational Titles* (DOT 007.161), Superintendent of Documents, U.S. Government Printing Office, Washington, D.C. 20402, 1978.

Fig. 10-21   Hydraulic component testing (*Courtesy of Commercial Shearing, Inc.*)

such procedures are often proposed and adopted by standards organizations, such as the American Society for Testing and Materials (ASTM) or the International Standards Organization (ISO), for general use.

Whenever possible, both the test setup and the reporting procedure should follow a standard format. The job sheet is suggested as the method for assembling equipment, diagnosing failures, making routine inspections, and repairing machinery components. Often, a procedure or job sheet will not be available, particularly where the test setup is novel or peculiar to a particular product or laboratory. In this case one should be constructed for present and future use, as well as for reporting the work completed. The reporting procedure followed by the technician for the entire test is called a technical report, and is used to both guide the procedure and report the results of a specific test The following sections are commonly found in the technical report:

1. Title
2. Name of the technician and date of the test
3. Purpose or scope of the test
4. Summary of the test method
5. Equipment or apparatus used in the test setup
6. Safety precautions
7. Procedure
8. Data and calculations
9. Tables, graphs of data, and results
10. Conclusions
11. Appendices (optional)
12. Summary notes

Item 5 in the technical report describes the equipment or apparatus used in the test setup. Often this is the step that takes the most time to complete because of assembly or construction difficulties, and this is where the job sheet should be used to guide, or developed to report, pertinent work. Although job sheets are less formalized than the technical report, it must be remembered that they are to be used by a skilled person who will have facility with hand tools and specialized testing equipment such as pressure gauges to perform the work specified. Commonly the job sheet includes:

1. Name of the job, test, or work to be performed
2. Name of the technician or mechanic and the dates during which the work was completed
3. Specialized tools and equipment
4. Safety precautions
5. Procedure

Figure 10-22 shows a technician assembling components of a hydraulic system from a blueprint and job sheet to fabricate the test setup.

The actual format of both the technical report and job sheet places specific sections of the report on certain pages of the report. For example, the title, name of the technician, and dates of the test are usually placed on the first page. If they are not lengthy, the purpose and scope of the test, together with the summary of the test method, are placed on the second

Fig. 10-22 Technician assembling hydraulic system components (*Courtesy of Commercial Shearing, Inc.*)

page of the report, section numbers at the margin being used to designate their order. If these and succeeding sections of the report are lengthy, then a table of contents and tables, graphs, etc., should follow the title page, and succeeding pages should be numbered for easy location of desired sections.

The purpose or scope of the test identifies what it is intended to do. For example, it may provide useful data about the sealing capabilities of dynamic seals under certain conditions commonly found in industry, or it may have common usage in predicting the life of a part or component. It could also be that this test determines the operating characteristics, such as power, speed, flow rate, etc., of such components as pumps, motors, fans, impeller blades, or turbine runners. The importance of the purpose is that it identifies very early in the report what it is that was to be accomplished and may even describe what are common outcomes so that the appropriateness to the reader's need is addressed.

The summary that follows the purpose gives a statement of the outcomes of the test in one or two paragraphs. It provides the information needed by the reader both to guide thinking during the actual reading of the report, and to provide a quick reference later should the results of the test be forgotten. If it adds to the clarity of the summary, it is permissible to make several definitive statements in sequence that list the important results established by the test.

The equipment or apparatus used in the test setup should be described in sufficient detail that the experiment or test could be replicated by someone at a later time. It should include blueprints, photographs, sketches, a list of components with specifications and serial numbers, and circuits indicating the direction of fluid flow. Standard pictorial or circuit symbols are used for this purpose. Functions that may be peculiar to a circuit should be identified, particularly if these generated the reason for conducting the test. Such would be the case if a certain component or part were prone to malfunction during an operating cycle. When the equipment is familiar to the reader, simple listings in tabular form with necessary specifications will suffice, leaving sketches and blueprints clear of unnecessary data.

Safety precautions are listed to ensure that the likelihood of accident, injury to personnel, and damage to equipment is as small as possible. Particularly, safety equipment such as eye protection, and procedural steps to protect personnel, such as tagging out an electrical circuit, must be emphasized, even though some of these may appear later as steps in the procedure. One way to ensure that important safety procedures are followed is to require a supervisor to sign off at specified intervals where these precautions should be taken. These signatures will then appear on the job sheet that is used to guide the test setup or on the technical report itself as one of the procedural steps.

The procedure lists the steps in sequence followed to conduct the test. It also notes important data that the test might yield, or specified settings of equipment which should be made or checked to ensure that each datum will be correct when it is collected. Pressures and flow rates, for example, should be checked before torque and horsepower readings of fluid pumps and motors are noted. To increase data accuracy, it is sometimes important to use repeated measures that are then averaged. The procedure should also emphasize if certain readings are to be held constant while others are allowed to vary, because this could affect the outcome of the test dramatically.

Data and calculations list the formulas used and the calculations that have been applied to the test in sufficient detail to allow a knowledgeable person to check the work. It is not necessary to derive formulas if a primary source is listed as a footnote or reference to corroborate the appropriateness of the formulas. Tables, graphs of data, and related results follow to show the reader the sequence of development of the outcomes of the test. That is, tables and graphs of data extend the raw data and calculations to give them meaning, either in pictorial form as displayed by graphs, or as positioned in a table that lists results of calculations.

Conclusions answer the question "so what." They allow the technician to interpret what the test means and what would likely occur if a product or material were used in an application with similar characteristics to the test procedure. Significance is attached to results and interpreted to the reader. Unusual occurrences or findings outside the scope of the test should also be reported if these impinge upon the intended purpose of the test or upon other related considerations. Difficulty of assembly of certain components would not have a direct bearing upon the performance of a component for example, but may have a significant effect on the use of the component as part of a larger system where ease of assembly was important in the construction or maintenance of equipment. A section for summary notes allows for future discussion or later consideration of items covered in the report. For example, when the report is presented to a supervisor or client, several questions about the test might surface, and the answers to these can then be recorded at the end of the report. Modifications to the report or to the procedure used to generate the report are also made from time to time, and these can be noted by the supervisor or technician for the time when the test will be repeated. Figure 10-23 shows how group effort is applied to solving technical problems, which might culminate either in generating a test procedure to establish a design solution, or in explaining the solution itself from a technical report that has been written after the test has been conducted to corroborate a proposed solution. Much effort and capital are expended by industry in this process, which generates many technical reports for future reference.

Fig. 10-23 Engineering group solving problems (*Courtesy of Commercial Shearing, Inc.*)

10-7 SUMMARY AND APPLICATIONS

Fluid machinery generates or consumes fluid power for useful purposes. Industrial and mobile applications generate power by using a prime mover to drive a pump, thus raising the pressure and energy level of the fluid. These prime movers are typically electric motors and internal combustion engines. In larger stationary installations, such as hydroelectric water projects, the potential energy of a water source is converted to electricity by driving a turbine, which is connected to a 50- or 60-cycle alternating current generator. Pump/turbine installations convert the potential energy of the water source to high-cost electricity during peak periods, and low-cost electricity to potential energy during low-use periods by returning the water to its original source or supplying water to irrigation projects at other locations at higher than pump elevations.

Positive displacement machinery processes a fixed volume of fluid for each cycle of the machine, for example, fluid power cylinders and most pumps and motors. Positive displacement machinery may have a fixed or variable volume, depending upon whether or not provision is made to vary the internal geometry of the component. Nonpositive displacement machinery typically converts the flow of fluid across runner vanes to power a shaft, or the converse, the spinning of an impeller to increase the velocity of the fluid to provide fluid power. Thus, turbo-machinery causes a flow rate with a static head to turn a shaft as the fluid passes through the runner, whereas pump counterparts first increase the velocity of the fluid across the rotating impeller and then through a defuser to create a flow rate with a static head. Generating power by passing the fluid through turbines is the more efficient of the two conversions.

Technicians in engineering and industrial-related occupations support the development, manufacture, installation, and maintenance of fluid machinery. A portion of their work is involved with testing and writing reports. Often they are called upon to make recommendations about specific products and procedures based upon experience and tests that have been conducted in the lab. Both a knowledge of the testing procedures and skills in reporting are necessary to their professional development.

Following are several applications that apply both testing and report-writing skills:

1. Construct a job sheet for a specific application such as installation, rebuilding, or testing a fluid power component (pump, motor, valve, hydrostatic transmission, etc.).
2. Test a centrifugal pump and write a technical report describing its operational characteristics and suitability for a specific application.
3. Write a historical sketch about one of the following: Pelton wheel turbine, Francis turbine, propeller turbine, Kaplan turbine.

4. Write a case study about a specific hydraulic turbine or pump/turbine installation (such as turbines 22, 23, and 24 at Grand Coulee Dam).

5. Report on a recent fluid power application by locating the product in a current trade journal and then consulting the manufacturer for literature and specifications.

6. Design a testing procedure for a specific fluid power component.

7. Report on specific certification requirements for one of the following: engineering technician, industrial technician, fluid power technician, fluid power mechanic.

10-8   STUDY QUESTIONS AND PROBLEMS

1. A hydraulic jack with a ram area of 40 cm^2 (6.2 in^2) is designed to raise a mass of 100 MG (megagrams = kg × 10^3) (220 000 lbf). Neglecting friction, what area of the pump piston would be necessary to raise the load if it were moved with a force of 670 N (150 lbf), and what would be the ratio of the ram piston area to the pump piston area?

2. Compute the effective volumetric displacement of a pump that delivers 125 1/min (27.5 gpm) at 1500 rpm.

3. If the pump in Problem 2 operates at 13.8 MPa (138 bars) (2001 psi) and the specifications indicate that the displacement should deliver 136.4 1/min (30 gpm), compute the torque applied to the shaft and the volumetric efficiency for the pump.

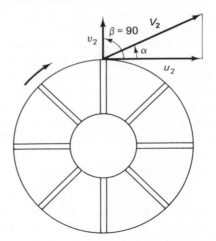

Fig. 10-24   Problem 6

4. From Problems 2 and 3, compute the volumetric losses from the pump as (a) actual internal leakage and (b) as a percentage of theoretical flow.

5. A centrifugal pump impeller with a diameter of 10.2 cm (4 in.) turns at 1725 rpm. Compute the angular velocity in radians/second.

6. A centrifugal pump impeller with a diameter of 12.7 cm (5 in.) and an angle $\beta=90$ deg has an exit velocity from the impeller of $v_2=10$ m/s (32.8 ft/s) when it turns at 1800 rpm. If $v_1=0$, compute the magnitude and direction of the resultant velocity vector V_2 (Fig. 10-24).

7. If the pump in Problem 6 delivers 135 l/min (29.7 gpm), compute the power. Assume overall efficiency (e_o) to be 60 percent.

8. Compute the delivery pressure for the pump in Problem 6.

9. What is the maximum horsepower that might be realized from water with a head of 30 m (98.43 ft) flowing at 4 m³/s (141.3 ft³/s)?

10. A Kaplan turbine with a runner outside diameter of 3 m (9.84 ft) and a boss diameter of 2 m (6.56 ft) runs full. Assuming that the blades occupy 15 percent of the volume in the annular ring between the tips of the blades and the hub, and the flow rate is 20 m³/s (706 ft³/s), compute the average water velocity downward through the turbine.

11. In Problem 10, if the runner turns at 75 rpm, $\beta_1=25$ deg and $\beta_2=17$ deg with the water exiting the blades with a final direction downward, compute the output horsepower of the unit (Fig. 10-25).

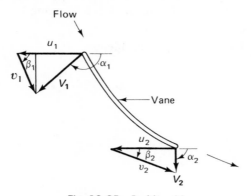

Flow

Fig. 10-25   Problem 11

12. Assuming no losses in Problem 11, what would be the head associated with the turbine?

13. A 90 000-hp Francis turbine operates at 100 rpm from a head of 78 m (256 ft). Compute the specific speed.

14. The Hoover Dam (Nevada) has two Francis turbines 3.35 m (11 ft) in diameter with a rated capacity of 115 000 hp at 180 rpm under a head of 146.3 m (480 ft). Compute the speed factor.

15. Compute the number of poles for the rotor of the alternator for the 700 MW units at Grand Coulee Dam if their speed of rotation is 85.7 rpm.

APPENDICES

A

CONVERSION FACTORS

$\text{bar} \times 10^5 = \text{Pa} \ (\text{N}/\text{m}^2)$

$\text{bar} = 14.5 \ \text{lbf}/\text{in}^2 \ (\text{approx. 1 atm.})$

$\text{dyne} \times 10^5 = \text{N}$

$\text{gal} \times (3.785 \times 10^{-3}) = \text{m}^3$

$\text{gal} \times 231 = \text{in}^3$

$\text{hp} \times 746 = \text{P (watts)}$

$\text{in} \times (2.54 \times 10^{-2}) = \text{m}$

$\text{lbf} \times 4.448 = \text{N}$

$\text{metric ton} \times 10^3 = \text{kgf}$

$\text{short ton (2000 lbf)} \times (9.072 \times 10^2) = \text{kgf}$

$\text{statute mile} \times (1.609 \times 10^3) = \text{m}$

$\gamma_{\text{water}} = 9802 \ \text{N}/\text{m}^3 = 62.4 \ \text{lbf}/\text{ft}^3 \ @ 10°\text{C} \ (50°\text{F})$

$\rho_{\text{water}} = 1000 \ \text{kg}/\text{m}^3 = 1.94 \ \text{slugs}/\text{ft}^3 \ @ 4°\text{C} \ (39.4°\text{F})$

ABSOLUTE VISCOSITY

$\text{lbf} \cdot \text{sec}/\text{ft}^2 \ (\text{no special name}) \times 47.88 = \text{Pa} \cdot \text{s} \ (\text{N} \cdot \text{s}/\text{m}^2)$

$\text{lbf} \cdot \text{sec}/\text{in}^2 \ (\text{rehns}) \times (6.895 \times 10^3) = \text{Pa} \cdot \text{s} \ (\text{N} \cdot \text{s}/\text{m}^2)$

$\text{Poise (dyne} \cdot \text{sec}/\text{cm}^2) \times 10^{-1} = \text{Pa} \cdot \text{s} \ (\text{N} \cdot \text{s}/\text{m}^2)$

$\text{Poise} \times 10^2 = \text{cP (Centipoise)}$

$\text{cP} \times 10^{-3} = \text{Pa} \cdot \text{s} \ (\text{N} \cdot \text{s}/\text{m}^2)$

KINEMATIC VISCOSITY

$\text{ft}^2/\text{sec} \ (\text{no special name}) \times (9.29 \times 10^{-2}) = \text{m}^2/\text{s}$

$\text{in}^2/\text{sec} \ (\text{Newts}) \times (6.45 \times 10^{-4}) = \text{m}^2/\text{s}$

$\text{Stoke (cm}^2/\text{s}) \times 10^{-4} = \text{m}^2/\text{s}$

$\text{Stoke} \times 10^2 = \text{cSt (Centistoke)}$

$$cSt \times 10^{-6} = m^2/s$$
$$cSt \times (15 \times 10^{-4}) = Newts$$

TRIGONOMETRIC FUNCTIONS

$$\sin\theta = \frac{y}{r}$$

$$\cos\theta = \frac{x}{r}$$

$$\tan\theta = \frac{y}{x}$$

$$\csc\theta = \frac{r}{y}$$

$$\sec\theta = \frac{r}{x}$$

$$\cot\theta = \frac{x}{y}$$

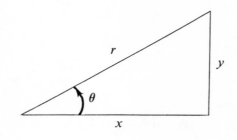

B

LETTER SYMBOLS AND ABBREVIATIONS

Letter Symbol	Quantity	SI Units	English Units	SI Symbol
A	Area	m^2, cm^2	ft^2, in^2	
C	Various constants			
C	Tensile strength	N/m^2	lbf/ft^2 lbf/in^2	Pa
D, d	Diameter	m, cm	ft, in.	
E	Specific energy	m	ft	
F	Force	N	lbf	N
H	Head energy	m	ft	
J	Joule constant		778 ft-lbf/Btu	
K, k	Various constants, factors			
L, l	Distance, length	m, cm	ft, in.	m
M	Mass	$kg(N \cdot s^2/m)$	$slug(lbf\text{-}sec^2/ft)$	M
N	Speed	rpm	rpm	
P	Power	J/s	J/s	
Q, q	Liquid volume flow rate	m^3/s, l/m	ft^3/sec, gal/min	
R	Hydraulic radius	m	ft	
S	Slope			
T	Torque	$N \cdot m$	lbf-ft, lbf-in.	
T	Temperature	°C	°F	
V	Volume	m^3	ft^3, in^3	
V	Vector velocity	m/s	ft/s	
W	Work	$m \cdot N$	ft-lbf	
a	Acceleration	m/s^2	ft/sec^2	
e	Efficiency			
f	Coefficient of friction			
g	Acceleration due to gravity	m/s^2	ft/sec^2	
h	Head	m	ft	
n	Revolutions			
n	Roughness factor	$s/m^{1/3}$	$s/ft^{1/3}$	

Letter Symbol	Quantity	SI Units	English Units	SI Symbol
p	Pressure	N/m^2	lbf/in^2 (psi)	Pa
			lbf/ft^2	
			bars	
r	Radius, hypotenuse	m, cm	ft, in.	
s	Percent slip			
s	Distance	m, cm	ft, in.	
t	Time	s	sec	s
u	Velocity	m/s	ft/sec	
u	Specific volume	m^3/kg	$ft^3/slug$	
v	Velocity	m/s	ft/sec	
w	Weight, load	N	lbf	
x	Distance	m	ft	
y	Flow depth	m	ft	
z	Potential energy from elevation	m	ft	

Greek Symbols	Quantity	SI Units	English Units	SI Symbol
α (alpha)	Angle			
β (beta)	Blade angle			
γ (gamma)	Specific weight	N/m^3	lbf/ft^3	
ϵ (epsilon)	Absolute roughness	m	ft	
θ (theta)	Angle			
μ (mu)	Absolute or dynamic viscosity	$N \cdot s/m^2$	$lbf \cdot sec/ft^2$ (rehns) $dyne \cdot sec/in^2$ (poise)	Pa·s
ν (nu)	Kinematic viscosity	m^2/s cm^2/s (stoke)	ft^2/sec in^2/sec (Newts)	
ϕ (phi)	Angle			
π (pi)	Constant			
ρ (rho)	Density	Kg/m^3	$slugs/ft^3$	
σ (sigma)	Surface tension	N/m	lbf/ft	
τ (tau)	Shear stress	N/m^2	lbf/ft^2	Ps
ω (omega)	Angular velocity	rad/s	rad/sec	

Other Symbols	Quantity	SI Units	English Units
FE	Flow-work energy	m-N	ft-lbf
KE	Kinetic energy	m-N	ft-lbf
PE	Potential energy	m-N	ft-lbf
c.p.	Center of pressure		
c.g.	Center of gravity		
w.p.	Wetted perimeter	m	ft

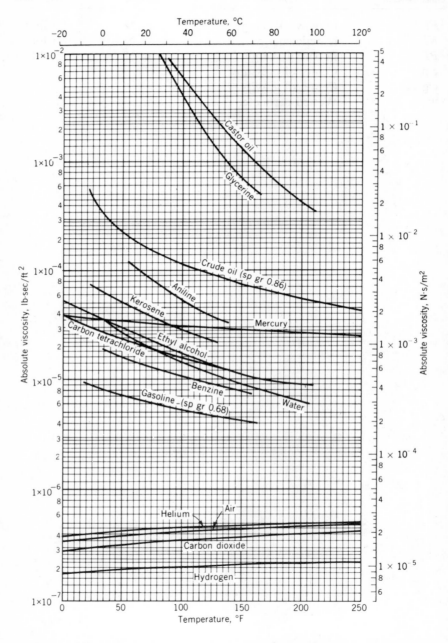

Absolute Viscosities of Several Gases and Liquids (*Reprinted with permission from Streeter and Wylie*, Fluid Mechanics, *Copyright 1975 by McGraw-Hill, Inc.*)

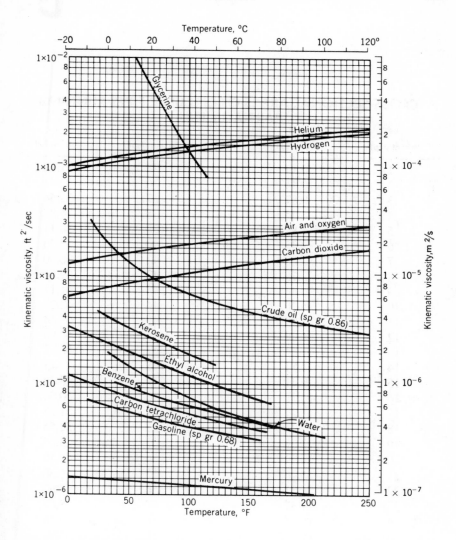

Kinematic Viscosities of Several Gases and Liquids (*Reprinted with permission from Streeter and Wylie,*** Fluid Mechanics, *Copyright 1975 by McGraw-Hill, Inc.***)**

D

CROSS-SECTIONAL AREAS OF CIRCULAR PIPES RUNNING PARTIALLY FULL

Percent depth	Percent full	Percent depth	Percent full
1	0.17	35	31.19
2	0.44	36	32.41
3	0.87	37	33.64
4	1.34	38	34.87
5	1.87	39	36.11
6	2.45	40	37.36
7	3.08	41	38.60
8	3.75	42	39.86
9	4.46	43	41.11
10	5.20	44	42.37
11	5.99	45	43.64
12	6.80	46	44.90
13	7.64	47	46.17
14	8.51	48	47.45
15	9.41	49	48.72
16	10.33	50	50.00
17	11.27	51	51.28
18	12.24	52	52.55
19	13.23	53	53.83
20	14.24	54	55.10
21	15.26	55	56.36
22	16.31	56	57.63
23	17.37	57	58.89
24	18.45	58	60.14
25	19.55	59	61.40
26	20.66	60	62.64
27	21.79	61	63.89
28	22.92	62	65.13
29	24.07	63	66.36
30	25.23	64	67.59
31	26.40	65	68.81
32	27.59	66	70.02
33	28.78	67	71.22
34	29.98	68	72.41

Percent depth	Percent full	Percent depth	Percent full
69	73.60	85	90.59
70	74.77	86	91.49
71	75.93	87	92.36
72	77.08	88	93.20
73	78.21	89	94.01
74	79.34	90	94.80
75	80.45	91	95.54
76	81.55	92	96.25
77	82.63	93	96.92
78	83.69	94	97.55
79	84.74	95	98.13
80	85.76	96	98.66
81	86.77	97	99.13
82	87.76	98	99.52
83	88.73	99	99.83
84	89.67	100	100.00

NATURAL TRIGONOMETRIC FUNCTIONS

Degrees	Sin	Cos	Tan	Cot	Sec	Cosec	
0	.0000	1.0000	.0000		1.0000		90
1	.0175	.9998	.0175	57.29	1.0000	57.30	89
2	.0349	.9994	.0349	28.64	1.001	28.65	88
3	.0523	.9986	.0524	19.08	1.001	19.11	87
4	.0698	.9976	.0699	14.30	1.002	14.34	86
5	.0872	.9962	.0875	11.43	1.004	11.47	85
6	.1045	.9945	.1051	9.514	1.006	9.567	84
7	.1219	.9925	.1228	8.144	1.008	8.206	83
8	.1392	.9903	.1405	7.115	1.010	7.185	82
9	.1564	.9877	.1584	6.314	1.012	6.392	81
10	.1736	.9848	.1763	5.671	1.015	5.759	80
11	.1908	.9816	.1944	5.145	1.019	5.241	79
12	.2079	.9781	.2126	4.705	1.022	4.810	78
13	.2250	.9744	.2309	4.331	1.026	4.445	77
14	.2419	.9703	.2493	4.011	1.031	4.134	76
15	.2588	.9659	.2679	3.732	1.035	3.864	75
16	.2756	.9613	.2867	3.487	1.040	3.628	74
17	.2924	.9563	.3057	3.271	1.046	3.420	73
18	.3090	.9511	.3249	3.078	1.051	3.236	72
19	.3256	.9455	.3443	2.904	1.058	3.072	71
20	.3420	.9397	.3640	2.747	1.064	2.924	70
21	.3584	.9336	.3839	2.605	1.071	2.790	69
22	.3746	.9272	.4040	2.475	1.079	2.669	68
23	.3907	.9205	.4245	2.356	1.086	2.559	67
24	.4067	.9135	.4452	2.246	1.095	2.459	66
25	.4226	.9063	.4663	2.145	1.103	2.366	65
26	.4384	.8988	.4877	2.050	1.113	2.281	64
27	.4540	.8910	.5095	1.963	1.122	2.203	63
28	.4695	.8829	.5317	1.881	1.133	2.130	62
29	.4848	.8746	.5543	1.804	1.143	2.063	61
30	.5000	.8660	.5774	1.732	1.155	2.000	60
31	.5150	.8572	.6009	1.664	1.167	1.942	59
32	.5299	.8480	.6249	1.600	1.179	1.887	58
	Cos	Sin	Cot	Tan	Coses	Sec	Degrees

Degrees	Sin	Cos	Tan	Cot	Sec	Cosec	
33	.5446	.8387	.6494	1.540	1.192	1.836	57
34	.5592	.8290	.6745	1.483	1.206	1.788	56
35	.5736	.8192	.7002	1.428	1.221	1.743	55
36	.5878	.8090	.7265	1.376	1.236	1.701	54
37	.6018	.7986	.7536	1.327	1.252	1.662	53
38	.6157	.7880	.7813	1.280	1.269	1.624	52
39	.6293	.7771	.8098	1.235	1.287	1.589	51
40	.6428	.7660	.8391	1.192	1.305	1.556	50
41	.6561	.7547	.8693	1.150	1.325	1.524	49
42	.6691	.7431	.9004	1.111	1.346	1.494	48
43	.6820	.7314	.9325	1.072	1.367	1.466	47
44	.6947	.7193	.9657	1.036	1.390	1.440	46
45	.7071	.7071	1.0000	1.000	1.414	1.414	45
	Cos	Sin	Cot	Tan	Cosec	Sec	Degrees

F

ANSWERS TO EVEN-NUMBERED PROBLEMS

Chapter 1

6. a. 290.8×10^4
   b. 141.9×10^{12}
   c. 1.7×10^7
   d. 0.5×10^6
   e. 14.6×10^{17}
8. a. 65 432, 6543, 654
   b. 17 654, 1765, 177
   c. 12 346, 1235, 124
   d. 21 378, 2138, 214
   e. 47 926, 4793, 479
10. a. 715 N
    b. 98 N
    c. 0.147 N
12. $P = KML^2T^{-3}$
14. $K = L^{1/2}M^{-1/2}$

Chapter 2

4. 1.53 kg/l
6. 8928.7 N/m^3
8. 0.87
10. $\mu = FL^{-2}T = ML^{-1}T$
12. 15.24 s
14. 18.17 cSt, 0.028 Newts

Chapter 3

4. Diameter Tabled values are force in newtons (N)

5.08 cm	5589	11 178	16 768	22 357	27 946
10.16 cm	22 357	44 714	67 070	89 427	111 784
15.24 cm	50 303	100 605	150 908	201 211	251 513
20.32 cm	89 427	178 854	268 281	357 708	447 135
25.40 cm	139 730	279 459	419 189	558 919	698 649
	2.759 MPa	5.517 MPa	8.276 MPa	11.034 MPa	13.793 MPa

Pressure (MPa)

6. Extension force $= 1.33$ retraction force

8. 19.34 mm (0.761 in.)

10. 3.26 in. (82.7 mm)

12. 51 132 N/pin (11 496 lbf/pin), $x_p = 0.507$ m (1.66 ft)

14. 5271 N (1185 lbf)

Chapter 4

6. 0.01 1/cyl/rev (0.61 in^3/cyl/rev)

8. 1.7 m/s (5.57 ft/sec)

12. $v_1 = 0.47$ m/s (1.54 ft/sec)
  $v_2 = 4.23$ m/s (13.88 ft/sec)

14. $v_1 = 0.64$ m/s (2.1 ft/s)
  $v_2 = 1.77$ m/s (5.80 ft/s)
  $p_2 = 890$ kPa (129 psi)

16. t (horizontal stream) $= 3.2$ s
  t (vertical stream) $= 1.69$ s

Chapter 5

2.

F	(kN)	12.225	12.225	12.225	12.225	12.225	12.225
	(lbf)	2748	2748	2748	2748	2748	2748
L	(m)	2	2	2	2	2	2
	(ft)	6.56	6.56	6.56	6.56	6.56	6.56
t	(sec)	30	25	20	15	10	5
HP		1.09	1.31	1.64	2.18	3.28	6.55

4. 10.35 hp
6. 29.67 N·m
   21.88 lbf·ft
8. 137 936 hp
10. 703 210
12. Upper limit $= 1.47 \times 10^{-2}$ m/s (0.048 ft/sec)
   Lower limit $= 7.35 \times 10^{-3}$ m/s (0.024 ft/sec)
14. 1.02 m/s (3.45 ft/sec)

Chapter 6

2. $f\left(\dfrac{L}{D}\right) = K$
4. $K = 0.56$
6. $L = 36.54$ m (119.9 ft)
8. 103 m, 24.5 hp
10. 0.22 m^3/s (7.77 ft^3/sec)
12. 17.9 cm (7.05 in)
14. $Q_A = 4666$ 1/min
   $Q_B = 334$ 1/min

Chapter 7

2. $R = 15.74$
4. $b = 0.54$ m (1.77 ft)
6. $r = 1.82$ m (5.97 ft)
8. 3.84 m/s (12.6 ft/sec)
10. $E = 1.75$ m (5.74 ft)
12. $Q = 70.8$ m^3/s (2500 ft^3/sec)
14. $S_c = 0.01$

Chapter 8

2. $Q = 0.135$ m^3/s (4.77 ft^3/sec)
   $-F = 202.5$ N (45.5 lbf)
4. $-F = 21.9$ N (4.93 lbf)
6. $R = 6.7$ m/s (22 ft/sec)
   $\theta = 35°$ approx.
8. $F_R = 235$ N (53 lbf)
   $\theta = 4°$ approx.

10. $-F_x = 417.6$ N (94 lbf)
    HP = 15.95
12. $F_x = p_1A - p_2A\cos\theta - \rho Qv(\cos\theta - 1)$
    $F_y = (p_2A + \rho Qv)(\sin\theta)$
14. $F_R = 9241$ N (2078 lbf)
    $\theta = 13°$ approx.

Chapter 9

2. $p = 1225$ Pa (0.18 lbf/in²)
   $f = 0.16$
4. 32.7 kPa (474 lbf/in²)
6. 0.11
8. 0.000 99
10. 0.0025
12. 75

Chapter 10

2. 83.3 cm³/rev (5.1 in³/rev)
4. 11.4 liters/min
   8.4 percent
6. $V_2 = 25.9$ m/s (85 ft/sec)
   $\alpha = 23°$ approx.
8. 5.71 bars (83 lbf/in²)
10. 6.0 m/s (19.7 ft)
12. 13.5 m (14.2 ft)
14. 0.59

INDEX